The Romance
of Modern Invention

Trending Technology in 1902

by
Archibald Williams

This edition published 2015 by Bosko Books.

Copyright 2015 Bosko Books, Business Unit, Manor Cottage, Church Road, Bristol, BS39 4EX

All trademarks used herein are the property of their respective owners. The use of any trademarks in this text does not vest in the author, editor or publisher any trademark ownership rights in such trademarks, nor does the use of such trademarks imply any affiliation with or endorsement of this book by such owners.

ISBN 0-9547239-7-X

A CIP catalogue record for this book is available from the British Library.

Typeset in Times New Roman, with headings and titles set in Perpetua with an 80% tint.

Bosko Books (www.boskobooks.com)

Bosko Books publishes books about people, technology and design. Both in the real world and in the digital world. From user experience design to technological innovation. From programming languages designed for users to public spaces designed for users. We can be contacted through the above web site.

The Romance
of Modern Invention

Contents

Section Two
Transport

Section Three
Visual Reproduction

Foreword

The text in this book is an abridged version of 'The Romance of Modern Invention' by Archibald Williams. It is the voice of a keen observer of technological advances writing in 1902. To some extent this period was the crucible of the information era. Although it was long before the advent of digital technology, new technologies were emerging and the wheels of industry were turning fast so that everybody was trying to be an inventor. Many modern devices and services that have entered our lives in the past few decades actually had their invention at the turn of the century in this flurry of technological innovation; familiar things like voice-mail, broadcast advertising, hybrid cars. This book even discusses mobile phones and 'couch-potato' syndrome!

Archibald's Style of Writing

The style of writing is very much of the turn of the last century, it can tend to be verbose and sometimes states the obvious. There are also occasional, throw-away comments about other nationalities of the sort that would cause a degree of consternation in today's, politically correct climate, for example the glib; 'Even Norway has a telephone system…', or; 'Roads, in places as bad as only German roads can be…'.

However, the writing style does not render the text impossible to follow, indeed it imparts upon it a sense of quaintness in keeping with the content, and the occasional nationalistic comment must be taken with a pinch of salt as they add a bit of spice to the proceedings.

Naming Things

These days the naming of products and services is a world unto itself; expensive experts cluster into groups to mull over new words and new interpretations of old words. They put focus groups together and spend ages interviewing people about their deepest associations with the words.

Back when Archibald was writing, the naming was done by the engineers, and as such it was structured and descriptive, if a trifle repetitive. Our word 'telephone' was created by combining the Greek words for 'distant' and 'sound' and in a similar way almost every other new invention was named by combining words in Greek and sometimes Latin. Thus, as well as the familiar telephone, this book will cover the phonograph, the teleautograph the telephonograph and more. It is a far cry from 'laptop', 'Wi-Fi', 'Google' and 'iPod'.

Units of Measurement

As the book was written in England in 1902 it uses the units of that country at that time. The unit of currency was the pound, amounts are expressed in pounds and shillings with 20 shillings making up one pound. To add confusion, the unit of weight was also called a pound, with one pound equivalent to 0.454 Kilograms. Thus a weight of two pounds is just under a kilogram. Finally, temperature was measured in degrees Fahrenheit rather than Centigrade.

Re-editing the Book's Structure

In re-releasing this book, the decision was taken to maintain the style of the text but to edit the coverage. With the amazing hindsight that a century gives us, we have made some changes. We have left out several sections in order to concentrate on the breakthroughs in communication and information technology and we have re-ordered the remaining chapters to give a structure more relevant to today's technological world. We have however left a few sections in that are not wholly relevant but do serve to give a flavour of the technologies in other areas at that particular moment in time.

For this reason the chapters on military advances have been dropped; tanks, dirigible torpedoes etc. However, there are still a few military references in the chapters on flying and automobiles, in particular in the chapter on 'Horseless Carriages'.

We have also skipped the chapters that dealt with wonders of the time, not so much great technological advances, but crowd-pulling examples of existing technology. Thus the chapter on the Great Paris Telescope is not included. Everyone knows about telescopes and the technology they embody, the Paris telescope was simply a very big example set on public display.

Finally there are miscellaneous sections that we decided to drop that covered areas outside the three sections below. These included liquid air, light bulb technology and solar motors. Everything that remained was gathered together under the following three sections:

Section 1 – Communications and Sound

This covers a variety of methods of communicating information at a distance, both through fixed wires and wireless. There are a number of different types of information that are dealt with, in particular; text, audio and graphics. This section also deals with technologies for the storage of sound for later playback.

Section 2 – Transport

This section considers the advances in the different modes of transport, namely; cars, trains, boats and planes. It also considers the different fuels that were being investigated at the start of the twentieth century.

Although not directly dealing with information or communication, the section has some very interesting insights into how the development of technology is influenced by cultural, political and other factors.

Section 3 – Visual Reproduction

Here the book deals with several technologies for reproducing moving pictures, a subject that in 1902 was just about to take off in a big way. Also covered are the setting of type for printing and early attempts at colour photography.

What to Expect in This Book

So what can the reader expect from this text? Some parts are heavy-going as they focus on the detailed workings of the technology, but the majority of the work is colourful and interesting. To offer something of a taster of the content here are a few highlights of the subjects covered:

Telefon Hirmondo (in chapter 3)

This was the name of the Hungarian telephone broadcast system at the turn of the century. Many people involved in the world of technology today know that Telefon Hirmondo existed, but Williams does more than just describe the technology and underlying idea. He discusses the service in some detail, the concerts and news broadcasts, outlining something akin to today's web services, it carried news and entertainment, it even had advertising, long before the internet or commercial radio advertising and he goes on to envisage that it will all lead to a telephone-based world (starting in Hungary of course) where everyone stays in bed all day and just does everything by telephone:

> 'If the principle is much further developed, we shall begin to doubt whether a Buda-Pesthian will be able to discover reasons for getting out of bed at all if the receiver hanging within reach of his hand is the entrance to so many places of delight.'

Innovation (in chapter 7)

Then there is a fine example of how new technology is initially born out of old technologies. The new technology in question is the car or the 'horseless carriage' and two of the first attempts at solving the problem were conducted by experts in old technologies applying their technologies to this new area. Firstly we have Simon Stevin; an enterprising character from Holland, the land of wind, windmills and sail boats, his attempts at the horseless carriage involved fitting carriages with huge sails. Next was Johann Haustach from Nuremberg, a watchmaker by trade, who applied technology from his world to the nascent automobile by hitting upon the idea of powering a car using a massive watch spring!

Dis-innovation (in chapter 7)

The rest of the chapter on cars gives an insight into the many factors that control the development of new technology. I have often heard that early motor cars by law had to have someone walking in front of them with a red flag to warn people and prevent accidents. According to Mister Williams this was indeed the case, but the reasoning he recounts is more out of politicking and profiteering than concern for the safety of road users.

The law was introduced by British members of parliament keen to protect their 'horsey' lifestyle and their investment in the railway companies. When a steam powered coach was able to outpace Stephenson's train 'The Rocket' in 1830, railway stakeholders lobbied parliament, pointed to the two road accidents that had so far taken place, and got them to pass the 'red flag law' in 1836, thus literally putting the brakes on the rapid automobile development of the time.

While the British government stuck with their horses and put the dampers on the development of the car, the French, being less horse oriented, were far more supportive of the development of the car and this is probably one of the reasons why France still has the giants of Peugeot, Renault and Citroen, and even today a part of the vocabulary of the car is derived from the original French words, terms like chauffeur, coupe, cabriolet, sedan, even the words car and garage have French origins.

I'm On the Train! (in chapter 1)

Or how about this little snippet on the subject of a global communication network:

> 'If a person wished to call to a friend ... he would use a loud electromagnetic voice, audible only to him who had the electromagnetic ear'.

> 'Where are you?' he would say.

> 'The reply would come; 'I am at the bottom of a coal mine', or 'Crossing the Andes', or 'In the middle of the Pacific'. Or, perhaps, in spite of all the calling, no reply would come, and the person would then know his friend was dead.'

How familiar that is to commuters on trains over a hundred years later, where the chief role of the mobile phone seems to be less about communication and more about location. The only difference being that if you don't get a reply it's probably just a weak signal rather than your friend being dead.

... And More

And there is more than just foretelling the couch potato and the mobile phone, the book also comments on the invention and early trials of the monorail, Mr. Kumberg's telephonograph, or 'voicemail' as we now refer to it, and Pieper and Jenatsky's hybrid petrol/electric car.

Along the way we will also meet with David Gorton's steam carriage that propelled itself not with wheels but with metal legs, Captain Walker's unique experiments with the car in the theatre of war during the Boer War, a two hundred foot long sausage and the observation that banjo duets make the best recordings!

The Author

Archibald Williams was a prolific author who popularised the science of the day with books about civil and structural engineering, invention, mechanics and exploration. He also wrote articles for magazines including Pearson's Magazine and The Strand. Many of his works were aimed at introducing the subjects to children. A graduate of Oxford University and a Fellow of the Royal Geographical Society he was born on 14 July 1871 and died on 14 July 1934; his sixty-third birthday.

Lon Barfield

October 2015

Author's Preface

The object of this book is to set before young people in a bright and interesting way, and without the use of technical language, accounts of some of the latest phases of modern invention; and also to introduce them to recent discoveries of which the full development is yet to be witnessed.

The author gratefully acknowledges the help given him as regards both literary matter and illustrations by: Mr. Cuthbert Hall (the Marconi Wireless Telegraphy Company), Mr. William Sugg, Mr. Hans Knudsen, Mr. F. C. B. Cole, Mr. E. J. Ryves, Mr. Anton Pollak, the Telautograph Company, the Parsons Steam Turbine Company, the Monotype Company, the Biograph Company, the Locomobile Company and the Speedwell Motor Company.

Archibald Williams

September 1902

Section One

Communications and Sound

Chapter 1

Wireless Telegraphy

One day in 1845 a man named Tawell, dressed as a Quaker, stepped into a train at Slough Station on the Great Western Railway, and travelled to London. When he arrived in London the innocent-looking Quaker was arrested, much to his amazement and dismay, on the charge of having committed a foul murder in the neigh-bourhood of Slough. The news of the murder and a description of the murderer had been telegraphed from that place to Paddington, where a detective met the train and shadowed the miscreant until a convenient opportunity for arresting him occurred. Tawell was tried, condemned, and hung, and the public for the first time gener-ally realised the power for good dormant in the as yet little devel-oped electric telegraph.

Thirteen years later two vessels met in mid-Atlantic laden with cables which they joined and paid out in opposite directions, till Ireland and Newfoundland were reached. The first electric mes-sage passed on August 7th of that year from the New World to the Old. The telegraph had now become a world power.

The third epoch-making event in its history is of recent date. On December 12th, 1901, Guglielmo Marconi, a young Italian, famous all over the world when but twenty-two years old, suddenly sprang into yet greater fame. At Hospital Point, Newfoundland, he heard by means of a kite, a long wire, a delicate tube full of tiny particles of metal, and a telephone ear-piece, signals transmitted from far-off Cornwall by his colleagues. No wires connected Poldhu, the Cornish station, and Hospital Point. The three short dot signals, which in the Morse code signify the letter S, had been borne from place to place by the limitless, mysterious ether, that strange substance of which we now hear so much, of which wise men declare we know so little.

Marconi's great achievement, which was of immense importance, naturally astonished the world. Of course, there were not wanting those who discredited the report. Others, on the contrary, were seized with panic and showed their readiness to believe that the Atlantic had been spanned aerially, by selling off their shares in cable companies. To use the language of the money-market, there was a temporary 'slump' in cable shares. The world again woke up, this time to the fact that experiments of which it had heard faintly had at last culminated in a great triumph, marvellous in itself, and yet probably nothing in comparison with the revolution in the transmission of news that it heralded.

The subject of Wireless Telegraphy is so wide that to treat it fully in the compass of a single chapter is impossible. At the same time it would be equally impossible to pass it over in a book written with the object of presenting to the reader the latest developments of scientific research. Indeed, the attention that it has justly attracted entitle it, not merely to a place, but to a leading place; and for this reason these first pages will be devoted to a short account of the history and theory of Wireless Telegraphy, with some mention of the different systems by which signals have been sent through space.

On casting about for a point at which to begin, the writer is tempted to attack the great topic of the ether, to which experimenters in many branches of science are now devoting more and more attention, hoping to find in it an explanation of and connection between many phenomena which at present are of uncertain origin.

The Ether

What is ether? In the first place, its very existence is merely assumed, like that of the atom and the molecule. Nobody can say that he has actually seen or had any experience of it. The assumption that there is such a thing is justified only in so far as that assumption explains and reconciles phenomena of which we have experience, and enables us to form theories which can be scientifically demonstrated correct. What scientists now say is this: that everything which we see and touch, the air, the infinity of space itself, is permeated by a something, so subtle that, no matter how continuous a thing may seem, it is but a concourse of atoms separated by this something, the ether. Reasoning drove them to this conclusion.

It is obvious that an effect cannot come out of nothing. Put a clock under a bell-glass and you hear the ticking. Pump out the air and the ticking becomes inaudible. What is now not in the glass that was there before? The air. Reason, therefore, obliges us to conclude that air is the means whereby the ticking is audible to us. No air, no sound. Next, put a lighted candle on the further side of the exhausted bell-glass. We can see it clearly enough. The absence of air does not affect light. But can we believe that there is an absolute gap between us and the light? No! It is far easier to believe that the bell-glass is as full as the outside atmosphere of the something that communicates the sensation of light from the candle to the eye. Again, suppose we measure a bar of iron very carefully while cold and then heat it. We shall find that it has expanded a little. The iron atoms, we say, have become more energetic than before, repel each other and stand further apart. What then is in the intervening spaces? Not air, which cannot be forced through iron whether hot or cold. No! the ether which passes easily through crevices so small as to bar the way to the atoms of air.

Once more, suppose that to one end of our iron bar we apply the negative 'pole' of an electric battery, and to the other end the positive pole. We see that a current passes through the bar, whether hot or cold, which implies that it jumps across all the ether gaps, or rather is conveyed by them from one atom to another.

The conclusion then is that ether is not merely omnipresent, penetrating all things, but the medium whereby heat, light, electricity, perhaps even thought itself, are transmitted from one point to another.

In what manner is the transmission effected? We cannot imagine the ether behaving in a way void of all system.

The answer is, by a wave motion. The ether must be regarded as a very elastic solid. The agitation of a portion of it by what we call heat, light, or electricity, sets in motion adjoining particles, until they are moving from side to side, but not forwards; the resultant movement resembling that of a snake tethered by the tail.

These ether waves vary immensely in length. Their qualities and effects upon our bodies or sensitive instruments depend upon their length. By means of ingenious apparatus the lengths of various waves have been measured. When the waves number 500 billion per second, and are but the 40,000th of an inch long they affect our eyes and are named light; red light. At double the number and half the length, they give us the sensation of violet light.

When the number increases and the waves shorten further, our bodies are 'blind' to them; we have no sense to detect their presence. Similarly, a slower vibration than that of red light is imperceptible until we reach the comparatively slow pace of 100 vibrations per second, when we become aware of heat.

Ether waves may be compared to the notes on a piano, of which we are acquainted with some octaves only. The gaps, the unknown octaves, are being discovered slowly but surely. Thus, for example, the famous X-rays have been assigned to the topmost octave; electric waves to the notes between light and heat. Forty years ago Professor Clerk Maxwell suggested that light and electricity were very closely connected, probably differing only in their wavelength. His theory has been justified by subsequent research. The velocity of light (185,000 miles per second) and that of electric currents have been proved identical. Hertz, a professor in the university of Bonn, also showed (1887 – 1889) that the phenomena of light reflection, refraction, and concentration of rays-can be repeated with electric currents.

We therefore take the word of scientists that the origin of the phenomena called light and electricity is the same; vibration of ether. It at once occurs to the reader that their behaviour is so different that they might as well be considered of altogether different natures.

For instance, interpose the very thinnest sheet of metal between a candle and the eye, and the light is cut off. But the sheet will very readily convey electricity. On the contrary, glass, a substance that repels electricity, is transparent, i.e. gives passage to light. And again, electricity can be conveyed round as many corners as you please, whereas light will travel in straight lines only.

To clear away our doubts we have only to take the lighted candle and again hold up the metal screen. Light does not pass through, but heat does. Substitute for the metal a very thin tank filled with a solution of alum, and then light passes, but heat is cut off. So that heat and electricity both penetrate what is impenetrable to light; while light forces a passage securely barred against both electricity and heat. And we must remember that open space conveys all alike from the sun to the earth.

On meeting what we call solid matter, ether waves are influenced, not because ether is wanting in the solid matter, but because the presence of something else than ether affects the intervening ether itself. Consequently glass, to take an instance, so affects ether that a very rapid succession of waves (light) are able to continue their way through its interstices, whereas long electric waves are so hampered that they die out altogether. Metal on the other hand welcomes slow vibrations (i.e. long waves), but speedily kills the rapid shakes of light. In other words, transparency is not confined to light alone. All bodies are transparent to some variety of rays, and many bodies to several varieties. It may perhaps even be proved that there is no such thing as absolute resistance, and that our inability to detect penetration is due to lack of sufficiently delicate instruments. The cardinal points to be remembered are these:

That the ether is a universal medium, conveying all kinds and forms of energy.

That these forms of energy differ only in their rates of vibration.

That the rate of vibration determines what power of penetration the waves shall have through any given substance.

Now, it is generally true that whereas matter of any kind offers resistance to light, that is, is not so perfect a conductor as the ether, many substances, especially metals, are more sensitive than ether to heat and electricity. How quickly a spoon inserted into a hot cup of tea becomes uncomfortably hot, though the hand can be held very close to the liquid without feeling more than a gentle warmth. And we all have noticed that the very least air-gap in an electric circuit effectively breaks a current capable of traversing miles of wire. If the current is so intense that it insists on passing the gap, it leaps across with a report, making a spark that is at once intensely bright and hot. Metal wires are to electricity what speaking tubes are to sound; they are as it were electrical tubes through the air and ether. But just as a person listening outside a speaking tube might faintly hear the sounds passing through it, so an instrument gifted with an 'electric ear' would detect the currents passing through the wire. Wireless telegraphy is possible because mankind has discovered instruments which act as electric ears or eyes, catching and recording vibrations that had hitherto remained undetected.

The earliest known form of wireless telegraphy is transmission of messages by light. A man on a hill lights a lamp or a fire. This represents his instrument for agitating the ether into waves, which proceed straight ahead with incredible velocity until they reach the receiver, the eye of a man watching at a point from which the light is visible.

Then came electric telegraphy. At first a complete circuit (two wires) was used. But in 1838 it was discovered that if instead of two wires only one was used, the other being replaced by an earth connection, not only was the effect equally powerful, but even double of what it was with the metallic circuit.

Thus the first step had been taken towards wireless electrical telegraphy. The second was, of course, to abolish the other wire. This was first effected by Professor Morse, who, in 1842, sent signals across the Susquehanna River without metallic connections of any sort. Along each bank of the river was stretched a wire three times as long as the river was broad. In the one wire a battery and transmitter were inserted, in the other a receiving instrument or galvanometer. Each wire terminated at each end in a large copper plate sunk in the water. Morse's conclusions were that provided the wires were long enough and the plates large enough, messages could be transmitted for an indefinite distance; the current passing from plate to plate, though a large portion of it would be lost in the water.

About the same date a Scotsman, James Bowman Lindsay of Dundee, a man as rich in intellectual attainments as he was pecuniarily poor, sent signals in a similar manner across the River Tay. In September, 1859, Lindsay read a paper before the British Association at Dundee, in which he maintained that his experiments and calculations assured him that by running wires along the coasts of America and Great Britain, by using a battery having an acting surface of 130 square feet and immersed sheets of 3,000 square feet, and a coil weighing 300 pounds, he could send messages from Britain to America.

Want of money prevented the poor scholar of Dundee from carrying out his experiments on a large enough scale to obtain public support. He died in 1862, leaving behind him the reputation of a man who in the face of the greatest difficulties made extraordinary electrical discoveries at the cost of unceasing labour; and this in spite of the fact that he had undertaken and partly executed a gigantic dictionary in fifty different languages!

The transmission of electrical signals through matter, metal, earth, or water, is effected by conduction, or the leading of the currents in a circuit. When we come to deal with aerial transmission, i.e. where one or both wires are replaced by the ether, then two methods are possible, those of induction and Hertzian waves.

Wireless Telegraphy by Induction

To take the induction method first. Whenever a current is sent through a wire, magnetism is set up in the ether surrounding the wire, which becomes the core of a 'magnetic field'. The magnetic waves extend for an indefinite distance on all sides, and on meeting a wire parallel to the electrified wire induce in it a dynamic current similar to that which caused them. Wherever electricity is present there is magnetism also, and vice versa. Electricity, produces magnetism, produces electricity. The invention of the Bell telephone enabled telegraphers to take advantage of this law.

In 1885 Sir William Preece, now consulting electrical engineer to the General Post-Office, erected near Newcastle two insulated squares of wire, each side 440 yards long. The squares were horizontal, parallel, and a quarter of a mile apart. On currents being sent through the one, currents were detected in the other by means of a telephone, which remained active even when the squares were separated by 1,000 yards. Sir William Preece thus demonstrated that signals could be sent without even an earth connection, i.e. entirely through the ether. In 1886 he sent signals between two parallel telegraph wires 41 miles apart. And in 1892 established a regular communication between Flatholm, an island fort in the Bristol Channel, and Lavernock, a point on the Welsh coast over three miles distant.

The inductive method might have attained to greater successes had not a formidable rival appeared in the Hertzian waves.

Wireless Telegraphy Equipment

In 1887 Professor Hertz discovered that if the discharge from a Leyden jar were passed through wires containing an air-gap across which the discharge had to pass, sparks would also pass across a gap in an almost complete circle or square of wire held at some distance from the jar. This 'electric eye', or detector, could have its gap so regulated by means of a screw that at a certain width its effect would be most pronounced, under which condition the detector, or receiver, was 'in tune' with the exciter, or transmitter. Hertz thus established three great facts, that:

A discharge of static (i.e. collected) electricity across an air-gap produced strong electric waves in the ether on all sides.

That these waves could be caught.

That under certain conditions the catcher worked most effectively.

Out of these three discoveries has sprung the latest phase of wireless telegraphy, as exploited by Signor Marconi. He, in common with Professors Branly of Paris, Popoff of Cronstadt, and Slaby of Charlottenburg, besides many others, have devoted their attention to the production of improved means of sending and receiving the Hertzian waves. Their experiments have shown that two things are required in wireless telegraphy:

That the waves shall have great penetrating power, so as to pierce any obstacle.

That they shall retain their energy, so that a maximum of their original force shall reach the receiver.

The first condition is fulfilled best by waves of great length; the second by those which, like light, are of greatest frequency. For best telegraphic results a compromise must be effected between these extremes, neither the thousand-mile long waves of an alternating dynamo nor the light waves of many thousands to an inch being of use. The Hertzian waves are estimated to be 230,000,000 per second; at which rate they would be 1.5 yards long. They vary considerably, however, on both sides of this rate and dimension.

Marconi's transmitter consists of three parts: a battery; an induction coil, terminating in a pair of brass balls, one on each side of the air-gap; and a Morse transmitting-key. Upon the key being depressed, a current from the battery passes through the coil and accumulates electricity on the brass balls until its tension causes it to leap from one to the other many millions of times in what is called a spark. The longer the air-gap the greater must be the accumulation before the leap takes place, and the greater the power of the vibrations set up. Marconi found that by connecting a kite or balloon covered with tinfoil by an aluminium wire with one of the balls, the effect of the waves was greatly increased. Sometimes he replaced the kite or balloon by a conductor placed on poles two or three hundred feet high, or by the mast of a ship.

We now turn to the receiver. In 1879 Professor D. E. Hughes observed that a microphone, in connection with a telephone, produced sounds in the latter even when the microphone was at a distance of several feet from coils through which a current was passing. A microphone, it may be explained, is in its simplest form a loose connection in an electric circuit, which causes the current to flow in fits and starts at very frequent intervals. He discovered that a metal microphone stuck, or cohered, after a wave had influenced it, but that a carbon microphone was self-restoring, i.e. regained its former position of loose contact as soon as a wave effect had ceased.

In 1891 Professor Branly of Paris produced a 'coherer', which was nothing more than a microphone under another name. Five years later Marconi somewhat altered Branly's contrivance, and took out a patent for a coherer of his own.

It is a tiny glass tube, about two inches long and a tenth of an inch in diameter inside. A wire enters it at each end, the wires terminating in two silver plugs fitting the bore of the tube. A space of $1/32$ inch is left between the plugs, and this space is filled with special filings, a mixture of 96 parts of nickel to 4 of silver, and the merest trace of mercury. The tube is exhausted of almost all its air before being sealed.

This little gap filled with filings is, except when struck by an electric wave, to all practical purposes a non-conductor of electricity. The metal particles touch each other so lightly that they offer great resistance to a current.

But when a Hertzian wave flying through the ether strikes the coherer, the particles suddenly press hard on one another, and make a bridge through which a current can pass. The current works a 'relay', or circuit through which a stronger current passes, opening and closing it as often as the coherer is influenced by a wave. The relay actuates a tapper that gently taps the tube after each wave-influence, causing the particles to decohere in readiness for the succeeding wave, and also a Morse instrument for recording words in dots and dashes on a long paper tape.

The coherer may be said to resemble an engine driver, and the 'relay' an engine. The driver is not sufficiently strong to himself move a train, but he has strength enough to turn on steam and make the engine do the work. The coherer is not suitable for use with currents of the intensity required to move a Morse recorder, but it easily switches a powerful current into another circuit.

Marconi's Experiments

Want of space forbids a detailed account of Marconi's successes with his improved instruments as he gradually increased the distance over which he sent signals through space.

In 1896 he came to England. That year he signalled from a room in the General Post Office to a station on the roof 100 yards distant. Shortly afterwards he covered 2 miles on Salisbury Plain.

In May, 1897, he sent signals from Lavernock Point to Flatholm, 3.33 miles. This success occurred at a critical time, for Sir W. Preece had already, as we have seen, bridged the same gap by his induction method, and for three days Marconi failed to accomplish the feat with his apparatus, so that it appeared as though the newer system were the less effective of the two. But by carrying the transmitting instrument on to the beach below the cliff on which it had been standing, and joining it by a wire to the pole already erected on the top of the cliff, Mr. Marconi, thanks to a happy inspiration, did just what was needed; he got a greater length of wire to send off his waves from. Communication was at once established with Flatholm, and on the next day with Brean Down, on the other side of the Bristol Channel, and 8.66 miles distant.

By the end of February, 1902 Mr. Marconi, during a voyage to America on the S.S. Philadelphia, remained in communication with Poldhu, Cornwall, until the vessel was 1,550 miles distant, receiving messages on a Morse recorder for any one acquainted with the code to read. Signals arrived for a further 500 miles, but owing to his instruments not being of sufficient strength, Mr. Marconi could not reply. When the transatlantic achievement was announced at the end of 1901, there was a tendency in some quarters to decry the whole system. The critics laid their fingers on two weak points.

In the first place, they said, the speed at which the messages could be transmitted was too slow to insure that the system would pay. Mr. Marconi replied that there had been a time when one word per minute was considered a good working rate across the Atlantic cable; whereas he had already sent twenty-two words per minute over very long distances. A further increase of speed was only a matter of time.

The second objection raised centred on the lack of secrecy resulting from signals being let loose into space to strike any instrument within their range; and also on the confusion that must arise when the ether was traversed by many sets of electric waves.

The young Italian inventor had been throughout his experiments aware of these defects and sought means to remedy them. In his earliest attempts we find him using parabolic metal screens to project his waves in any required direction and prevent their going in any other. He also employed strips of metal in conjunction with the coherer, the strips or 'wings' being of such a size as to respond most readily to waves of a certain length.

Tuned Wireless Telegraphy

The electric oscillations coming from the aerial wires carried on poles, kites, etc., were of great power, but their energy dispersed very quickly into space in a series of rapidly diminishing vibrations. This fact made them affect to a greater or less degree any receiver they might encounter on their wanderings. If you go into a room where there is a piano and make a loud noise near the instrument a jangle of notes results. But if you take a tuning-fork and after striking it place it near the strings, only one string will respond, i.e. that of the same pitch as the fork.

What is required in wireless telegraphy is a system corresponding to the use of the tuning-fork. Unfortunately, it has been discovered that the syntony or tuning of transmitter and receiver reduces the distance over which they are effective. An electric 'noise' is more far-reaching than an electric 'note'.

Mr. Marconi has, however, made considerable advances towards combining the sympathy and secrecy of the tuning system with the power of the 'noise' system. By means of delicately adjusted 'wings' and coils he has brought it about that a series of waves having small individual strength, but great regularity, shall produce on the receiver a cumulative effect, storing, as it were, electricity on the surface of the receiver 'wings' until it is of sufficient power to overcome the resistance of the coherer.

That tuned wireless telegraphy is, over moderate distances, at least as secret as that through wires (which can be tapped by induction) is evident from the fact that during the America Cup Yacht Races Mr. Marconi sent daily to the New York Herald messages of 4,000 total words, and kept them private in spite of all efforts to intercept them. He claims to have as many as 250 'tunes' and, indeed, there seems to be no limit to their number, so that the would-be 'tapper' is in the position of a man trying to open a letter-lock of which he does not know the cipher-word. He may discover the right tune, but the chances are greatly against him. We may be certain that the rapid advance in wireless telegraphy will not proceed much further before syntonic messages can be transmitted over hundreds if not thousands of miles.

It is hardly necessary to dwell upon the great prospect that the new telegraphy opens to mankind. The advantages arising out of a ready means of communication, freed from the shackles of expensive connecting wires and cables are, in the main, obvious enough. We have only to imagine all the present network of wires replaced or supplemented by etherwaves, which will be able to act between points (e.g. ships and ships, ships and land, moving and fixed objects generally) which cannot be connected by metallic circuits.

Already ocean voyages are being shortened as regards the time during which passengers are out of contact with the doings of the world. The transatlantic journey has now a newsless period of but three days. Navies are being fitted out with instruments that may play as important a part as the big guns themselves in the next naval war. A great maritime nation like our own should be especially thankful that the day is not far distant when our great empire will be connected by invisible electric links that no enemy may discover and cut.

The romantic side of wireless telegraphy has been admirably touched in some words uttered by Professor Ayrton in 1899, after the reading of a paper by Mr. Marconi before the Institution of Electrical Engineers.

'If a person wished to call to a friend' (said the Professor), 'he would use a loud electromagnetic voice, audible only to him who had the electromagnetic ear'.

'Where are you?' he would say.

'The reply would come; 'I am at the bottom of a coal mine', or 'Crossing the Andes', or 'In the middle of the Pacific'. Or, perhaps, in spite of all the calling, no reply would come, and the person would then know his friend was dead. Let them think of what that meant; of the calling which went on every day from room to room of a house, and then imagine that calling extending from pole to pole; not a noisy babble, but a call audible to him who wanted to hear and absolutely silent to him who did not'.

When will Professor Ayrton's forecast come true? Who can say? Science is so full of surprises that the ordinary man wonders with a semi-fear what may be the next development; and wise men like Lord Kelvin humbly confess that in comparison with what has yet to be learnt about the mysterious inner workings of Nature their knowledge is but as ignorance.

Chapter 2

High Speed Telegraphy

The wonderful developments of wireless telegraphy must not make us forget that some very interesting and startling improvements have been made in connection with the ordinary wire-circuit method: notably in the matter of speed. At certain seasons of the year or under special circumstances which can scarcely be foreseen, a great rush takes place to transmit messages over the wires connecting important towns. Now, the best telegraphists can with difficulty keep up a transmitting speed of even fifty words a minute for so long as half-an-hour. The Morse alphabet contains on the average three signals for each letter, and the average length of a word is six letters. Fifty words would therefore contain between them 900 signals, or fifteen a second. The strain of sending or noting so many for even a brief period is very wearisome to the operator. Means have been found of replacing the telegraph clerk, so far as the actual signalling is concerned, by mechanical devices.

The Chemical Telegraph

In 1842 Alexander Bain, a watchmaker of Thurso, produced what is known as a 'chemical telegraph'. The words to be transmitted were set up in large metal type, all capitals, connected with the positive pole of a battery, the negative pole of which was connected

15

to earth. A metal brush, divided into five points, each terminating a wire, was passed over the metal type. As often as a division of the brush touched metal it completed the electric circuit in the wire to which it was joined, and sent a current to the receiving station, where a similar brush was passing at similar speed over a strip of paper soaked in iodide of potassium. The action of the electricity decomposed the solution, turning it blue or violet. The result was a series of letters divided longitudinally into five belts separated by white spaces representing the intervals between the contact points of the brush.

The Bain Chemical Telegraph was able to transmit the enormous number of 1,500 words per minute; that is, at ten times the rate of ordinary conversation! But even when improvements had reduced the line wires from five to one, the system, on account of the method of composing the message to be sent, was not found sufficiently practical to come into general use.

Its place was taken by slower but preferable systems: those of duplex and multiplex telegraphy.

Duplex and Multiplex Telegraphy

When a message is sent over the wires, the actual time of making the signals is more than is required for the current to pass from place to place. This fact has been utilised by the inventors of methods whereby two or more messages may not only be sent the same way along the same wire, but may also be sent in different directions. Messages are 'duplex' when they travel across one another, 'multiplex' when they travel together.

The principle whereby several instruments are able to use the same wire is that of distributing among the instruments the time during which they are in contact with the line.

Let us suppose that four transmitters are sending messages simultaneously from London to Edinburgh.

Wires from all four instruments are led into a circular contact-maker, divided into some hundreds of insulated segments connected in rotation with the four transmitters. Thus instrument A will be joined to segments 1, 5, 9, 13; instrument B to segments 2, 6, 10, 14; instrument C with segments 3, 7, 11, 15; and so on.

Along the top of the segments an arm, connected with the telegraph line to Edinburgh, revolves at a uniform rate. For about $1/500$ of a second it unites a segment with an instrument. If there are 150 segments on the 'distributor', and the arm revolves three times a second, each instrument will be put into contact with the line rather oftener than 110 times per second. And if the top speed of fifty words a minute is being worked to, each of the fifteen signals occurring in each second will be on the average divided among seven moments of contact.

A similar apparatus at Edinburgh receives the messages. It is evident that for the system to work satisfactorily, or even to escape dire confusion, the revolving arms must run at a level speed in perfect unison with one another. When the London arm is over segment 1, the Edinburgh arm must cover the same number. The greatest difficulty in multiplex telegraphy has been to adjust the timing exactly.

The Phonic Wheel

Paul la Cour of Copenhagen invented for driving the arms a device called the Phonic Wheel, as its action was regulated by the vibrations of a tuning fork. The wheel, made of soft iron, and toothed on its circumference, revolves at a short distance from the pole of a magnet. As often as a current enters the magnet the latter attracts the nearest tooth of the wheel; and if a regular series of currents pass through it the motion of the wheel will be uniform. M. la Cour produced the regularity of current impulses in the motor magnet by means of a tuning-fork, which is unable to vibrate more than a certain number of times a second, and at each vibration closed a circuit sending current into the magnet. To get two tuning-forks of the same note is an easy matter; and consequently a uniformity of rotation at both London and Edinburgh stations may be insured.

So sensitive is this 'interrupter' system that as many as sixteen messages can be sent simultaneously, which means that a single wire is conveying from 500 to 800 words a minute. We can easily understand the huge saving that results from such a system; the cost of instruments, interrupter, etc., being but small in proportion to that of a number of separate conductors.

The word-sending capacity of a line may be even further increased by the use of automatic transmitters able to work much faster in signal-making than the human brain and hand. Sir Charles Wheatstone's Automatic Transmitter has long been used in the Post-Office establishments.

The Wheatstone Automatic Transmitter

The messages to be sent are first of all punched on a long tape with three parallel rows of perforations. The central row is merely for guiding the tape through the transmitting machine. The positions of the holes in the two outside rows relatively to each other determine the character of the signal to be sent. Thus, when three holes (including the central one) are abreast, a Morse 'dot' is signified; when the left-hand hole is one place behind the right hand, a 'dash' will be telegraphed.

In the case of a long communication the matter is divided among a number of clerks operating punching machines. Half-a-dozen operators could between them punch holes representing 250 to 300 words a minute; and the transmitter is capable of despatching as many in the same time, while it has the additional advantage of being tireless.

The action of the transmitter is based upon the reversal of the direction or nature of current. The punched tape is passed between an oscillating lever, carrying two points, and plates connected with the two poles of the battery. As soon as a hole comes under a pin the pin drops through and makes a contact.

At the receiving end the wire is connected with a coil wound round the pole of a permanent bar magnet. Such a magnet has what is known as a north pole and a south pole, the one attractive and the other repulsive of steel or soft iron. Any bar of soft iron can be made temporarily into a magnet by twisting round it a few turns of a wire in circuit with the poles of a battery. But which will be the north and which the south pole depends on the direction of the current. If, then, a current passes in one direction round the north pole of a permanent magnet it will increase the magnet's attractive power, but will decrease it if sent in the other direction.

The 'dot' holes punched in the tape being abreast cause first a positive and then a negative current following at a very short interval; but the 'dash' holes not being opposite allow the positive current to occupy the wires for a longer period. Consequently the Morse marker rests for correspondingly unequal periods on the recording 'tape', giving out a series of dots and dashes, as the inker is snatched quickly or more leisurely from the paper.

The Pollak-Virag system

The Wheatstone recorder has been worked up to 400 words a minute, and when two machines are by the multiplex method acting together this rate is of course doubled.

As a speed machine it has, however, been completely put in the shade by a more recent invention of two Hungarian electricians, Anton Pollak and Josef Virag, which combines the perforated strip method of transmission with the telephone and photography. The message is sent off by means of a punched tape, and is recorded by means of a telephonic diaphragm and light marking a sensitised paper.

In 1898 the inventors made trials of their system for the benefit of the United Electrical Company of Budapest. The Hungarian capital was connected by two double lines of wire with a station 200 miles distant, where the two sets were joined so as to give a single circuit of 400 miles in length. A series of tests in all weathers showed that the Pollak-Virag system could transmit as many as 100,000 words an hour over that distance.

From Hungary the inventors went to the United States, in which country of 'records' no less than 155,000 words were despatched and received in the sixty minutes. This average; 2,580 words per minute, 43 per second, is truly remarkable! Even between New York and Chicago, separated by 950 odd miles, the wires kept up an average of 1,000 per minute.

The apparatus that produces these marvellous results is of two types. The one type records messages in the Morse alphabet, the other makes clearly-written longhand characters. The former is the faster of the two, but the legibility of the other more than compensates for the decrease of speed by one-half.

The Morse alphabet method closely resembles the Wheatstone system. The message is prepared for transmission by being punched on a tape. But there is this difference in the position of the holes, that whereas in the Wheatstone method two holes are used for each dot and dash, only one is required in the Pollak-Virag. If to the right of the central guiding line it signifies a 'dash' if to the left, a 'dot'.

The 'reversal-of-current' method, already explained, causes at the receiver end an increase or decrease in the power of a permanent magnet to attract or repel a diaphragm, the centre of which is connected by a very fine metal bar with the centre of a tiny mirror hinged at one side on two points. A very slight movement of the diaphragm produces an exaggerated movement of the mirror, which, as it tilts backwards and forwards, reflects the light from an electric lamp on to a lens, which concentrates the rays into a bright spot, and focuses them on to a surface of sensitised paper.

In their earliest apparatus the inventors attached the paper to the circumference of a vertical cylinder, which revolved at an even pace on an axle, furnished at the lower end with a screw thread, so that the portion of paper affected by the light occupied a spiral path from top to bottom of the cylinder.

In a later edition, however, an endless band of sensitised paper is employed, and the lamp is screened from the mirror by a horizontal mantle in which is cut a helical slit making one complete turn of the cylinder in its length. The mantle is rotated in unison with the machinery driving the sensitised band; and as it revolves, the spot at which the light from the filament can pass through the slit to the mirror is constantly shifting from right to left, and the point at which the reflected light from the mirror strikes the sensitised paper from left to right. At the moment when a line is finished, the right extremity of the mantle begins to pass light again, and the bright spot of light recommences its work at the left edge of the band, which has now moved on a space.

The movements of the mirror backwards and forwards produce on the paper a zigzag tracing known as syphon-writing. The record, which is continuous from side to side of the band, is a series of zigzag up-and-down strokes, corresponding to the dots and dashes of the Morse alphabet.

The apparatus for transmitting longhand characters is more complicated than that just described. Two telephones are now used, and the punched tape has in it five rows of perforations.

If we take a copy-book and examine the letters, we shall see that they all occupy one, two, or three bands of space. For instance, a, between the lines, occupies one band; g, two bands; and f, three. In forming letters, the movements of the fingers trace curves and straight lines, the curves being the resultants of combined horizontal and vertical movements.

Messrs. Pollak and Virag, in order to produce curves, were obliged to add a second telephone, furnished also with a metal bar joined to the mirror, which rests on three points instead of on two. One of these points is fixed, the other two represent the ends of the two diaphragm bars, which move the mirror vertically and horizontally respectively, either separately or simultaneously.

A word about the punched paper before going further. It contains, as we have said, five rows of perforations. The top three of these are concerned only with the up-and-down strokes of the letters, the bottom two with the cross strokes. When a hole of one set is acting in unison with a hole of the other set a composite movement or curve results.

The topmost row of all sends through the wires a negative current of known strength; this produces upward and return strokes in the upper zone of the letters: for instance, the upper part of a 't'. The second row passes positive currents of equal strength with the negative, and influences the up-and-down strokes of the centre zone, e.g. those of 'o'; the third row passes positive currents twice as strong as the negative, and is responsible for double-length vertical strokes in the centre and lower zones, e.g. the stroke in 'p'.

In order that the record shall not be a series of zigzags it is necessary that the return strokes in the vertical elements shall be on the same path as the out strokes; and as the point of light is continuously tending to move from left to right of the paper there must at times be present a counteracting tendency counterbalancing it exactly, so that the path of the light point is purely vertical. At other times not merely must the horizontal movements balance each other, but the right-to-left element must be stronger than the left-to-right, so that strokes such as the left curve of an e may be possible. To this end rows 4 and 5 of the perforations pass currents working the second telephone diaphragm, which moves the mirror on a vertical axis so that it reflects the ray horizontally.

It is said that by the aid of a special 'multiplex' device thirty sets of Pollak-Virag apparatus can be used simultaneously on a line! The reader will be able, by the aid of a small calculation, to arrive at some interesting figures as regards their united output.

Chapter 3
The Telephone

A common enough sight in any large town is a great sheaf of fine wires running across the streets and over the houses. If you traced their career in one direction you would find that they suddenly terminate, or rather combine into cables, and disappear into the recesses of a house, which is the Telephone Exchange. If you tracked them the other way your experience would be varied enough. Some wires would lead you into public institutions, some into offices, some into snug rooms in private houses. At one time your journey would end in the town, at another you would find yourself roaming far into the country, through green fields and leafy lanes until at last you ran the wire to earth in some large mansion standing in a lordly park. Perhaps you might have to travel hundreds of miles, having struck a 'trunk' line connecting two important cities; or you might even be called upon to turn fish and plunge beneath the sea for a while, groping your way along a submarine cable.

In addition to the visible overhead wires that traverse a town there are many led underground through special conduits. And many telephone wires never come out of doors at all, their object being to furnish communication between the rooms of the same house. The telephone and its friend, the electric-bell, are now a regular part of the equipment of any large premises. The master of the house goes to his telephone when he wishes to address the cook or the steward, or the head-gardener or the coachman. It saves time and labour.

Should he desire to speak to his town-offices he will, unless connected direct, 'ring up' the Exchange, into which, as we have seen, flow all the wires of the subscribers to the telephone system of that district. The ringing-up is usually done by rapidly turning a handle which works an electric magnet and rings a bell in the Exchange. The operator there, generally a girl, demands the number of the person with whom the ringer wants to speak, rings up that number, and connects the wires of the two parties.

In some exchanges, e.g. the new Post-Office telephone exchanges, the place of electric-bells is taken by lamps, to the great advantage of the operators, whose ears are thus freed from perpetual jangling. The action of unhooking the telephone receiver at the subscriber's end sends a current into a relay which closes the circuit of an electric lamp opposite the subscriber's number in the exchange. Similarly, when the conversation is completed the action of hanging up the receiver again lights another lamp of a different colour, given the exchange warning that the wires are free again.

In America, the country of automatic appliances, the operator is sometimes entirely dispensed with. A subscriber is able, by means of a mechanical contrivance, to put himself in communication with any other subscriber unless that subscriber is engaged, in which case a dial records the fact.

The Popularity of the Telephone

The popularity of the telephone may be judged from the fact that in 1901 the National Telephone Company's system transmitted over 807 millions of messages, as compared with 89 millions of telegrams sent over the Post Office wires. In America and Germany, however, the telephone is even more universally employed than in England. In the thinly populated prairies of West America the farm-houses are often connected with a central station many miles off, from which they receive news of the outer world and are able to keep in touch with one another. We are not, perhaps, as a nation sufficiently alive to the advantages of an efficient telephone system; and on this account many districts remain telephoneless because sufficient subscribers cannot be found to guarantee use of a system if established. It has been seriously urged that much of our country depopulation might be counteracted by a universal telephone service, which would enable people to live at a distance from the towns and yet be in close contact with them. At present, for the sake of convenience and ease of 'getting at' clients and customers, many business men prefer to have their homes just outside the towns where their business is. A cheap and efficient service open to every one would do away with a great deal of travelling that is necessary under existing circumstances, and by making it less important to live near a town allow people to return to the country.

Even Norway has a good telephone system. The telegraph is little used in the more thinly inhabited districts, but the telephone may be found in most unexpected places, in little villages hidden in the recesses of the fiords. Switzerland, another mountainous country, but very go-ahead in all electrical matters, is noted for the cheapness of its telephone services.

At Berne or Geneva a subscriber pays 4 pounds the first year, 2 pounds and 12 shillings the second year, and but one pound and twelve shillings the third. Contrast these charges with those of New York, where 15 pounds and 10 shillings to 49 pounds is levied annually according to service.

Hungary's Telefon Hirmondo

The telephone as a public benefactor is seen at its best at Buda-Pesth, the twin-capital of Hungary. In 1893, one Herr Theodore Buschgasch founded in that city a 'newspaper', if so it may be called, worked entirely on the telephone. The publishing office was a telephone exchange; the wires and instruments took the place of printed matter. The subscribers were to be informed entirely by ear of the news of the day.

The Telefon Hirmondo or 'Telephonic Newsteller', as the 'paper' was named, has more than six thousand subscribers, who enjoy their telephones for the very small payment of eighteen florins, or about a penny a day, for twelve hours a day.

News is collected at the central office in the usual journalistic way by telephone, telegraph, and reporters, it is printed by lithography on strips of paper six inches wide and two feet long. These strips are handed to 'stentors', or men with powerful and trained voices, who read the contents to transmitting instruments in the offices, whence it flies in all directions to the ears of the subscribers.

These last know exactly when to listen and what description of information they will hear, for each has over his receiver a pro-gramme which is rigidly adhered to. It must be explained at once that the Telefon Hirmondo is more than a mere newspaper, for it adds to its practical use as a first-class journal that of entertainer, lecturer, preacher, actor, political speaker, musician. The Telefon offices are connected by wire with the theatres, churches, and public halls, drawing from them by means of special receivers the sounds that are going on there, and transmitting them again over the wires to the thousands of subscribers. The Buda-Pesthian has therefore only to consult his programme to see when he will be in touch with his favourite actor or preacher. The ladies know just when to expect the latest hints about the fashions of the day. Nor are the children forgotten, for a special period is set aside weekly for their entertainment in the shape of lectures or concerts. The advertising fiend, too, must have his say, though he pays dearly for it. On payment of a florin the stentors will shout the virtues of his wares for a space of twelve seconds. The advertising periods are sandwiched in between items of news, so that the subscriber is bound to hear the advertisements unless he is willing to risk miss-ing some of the news if he hangs up his receiver until the 'puff' is finished.

Thanks to the Telefon Hirmondo the preacher, actor, or singer is obliged to calculate his popularity less by the condition of the seats in front of him than by the number of telephones in use while he is performing his part. On the other hand, the subscriber is spared a vast amount of walking, waiting, cab-hire, and expense generally. In fact, if the principle is much further developed, we shall begin to doubt whether a Buda-Pesthian will be able to discover reasons for getting out of bed at all if the receiver hanging within reach of his hand is the entrance to so many places of delight. Will he become a very lazy person; and what will be the effect on his entertainers when they find themselves facing benches that are used less every day? Will the sight of a row of telephone trumpets rouse the future Liddon, Patty, Irving, or Gladstone to excel themselves? It seems rather doubtful. Telephones cannot look interested or applaud.

Telephone Equipment

What is inside the simple-looking receiver that hangs on the wall beside a small mahogany case, or rests horizontally on a couple of crooks over the case? In the older type of instrument the transmitter and receiver are separate, the former fixed in front of the case, the latter, of course, movable so that it can be applied to the ear. But improved patterns have transmitter and receiver in a single movable handle, so shaped that the earpiece is by the ear while the mouthpiece curves round opposite the mouth. By pressing a small lever with the fingers the one or the other is brought into action when required.

The construction of the instrument, of which we are at first a little afraid, and with which we later on learn to become rather angry, is in its general lines simple enough. The first practical telephone, constructed in 1876 by Graham Bell, a Scotchman, consisted of a long wooden or ebonite handle down the centre of which ran a permanent bar-magnet, having at one end a small coil of fine insulated wire wound about it. The ends of the wire coil are led through the handles to two terminals for connection with the line wires. At a very short distance from the wire-wound pole of the magnet, firmly fixed by its edges, is a thin circular iron plate, covered by a funnel-shaped mouthpiece.

The iron plate is, when at rest, concave, its centre being attracted towards the pole of the magnet. When any one speaks into the mouthpiece the sound waves agitate the diaphragm (or plate), causing its centre to move inwards and outwards. The movements of the diaphragm affect the magnetism of the magnet, sometimes strengthening it, sometimes weakening it, and consequently exciting electric currents of varying strength in the wire coil. These currents passing through the line wires to a similar telephone excite the coil in it, and in turn affect the magnetism of the distant magnet, which attracts or releases the diaphragm near its pole, causing undulations of the air exactly resembling those set up by the speaker's words. To render the telephone powerful enough to make conversation possible over long distances it was found advisable to substitute for the one telephone a special transmitter, and to insert in the circuit a battery giving a much stronger current than could possibly be excited by the magnet in the telephone at the speaker's end.

Edison in 1877 invented a special transmitter made of carbon. He discovered that the harder two faces of carbon are pressed together the more readily will they allow current to pass; the reason probably being that the points of contact increase in number and afford more bridges for the current.

Accordingly his transmitter contains a small disc of lampblack (a form of carbon) connected to the diaphragm, and another carbon or platinum disc against which the first is driven with varying force by the vibrations of the voice.

The Edison transmitter is therefore in idea only a modification of the microphone. It acts as a regulator of current, in distinction to the Bell telephone, which is only an exciter of current. Modern forms of telephones unite the Edison transmitter with the Bell receiver.

The latter is extremely sensitive to electric currents, detecting them even when of the minutest power. We have seen that Marconi used a telephone in his famous transatlantic experiments to distinguish the signals sent from Cornwall. A telephone may be used with an 'earth return' instead of a second wire; but as this exposes it to stray currents by induction from other wires carried on the same poles or from the earth itself, it is now usual to use two wires, completing the metallic circuit. Even so a subscriber is liable to overhear conversations on wires neighbouring his own; the writer has lively recollections of first receiving news of the relief of Ladysmith in this manner.

Owing to the self-induction of wires in submarine cables and the consequent difficulty of forcing currents through them, the telephone is at present not used in connection with submarine lines of more than a very moderate length. England has, however, been connected with France by a telephone cable from St. Margaret's Bay to Sangatte, 23 miles; and Scotland with Ireland, Stranraer to Donaghadee, 26 miles. The former cable enables speech between London and Marseilles, a distance of 900 miles; and the latter makes it possible to speak from London to Dublin via Glasgow. The longest direct line in existence is that between New York and Chicago, the complete circuit of which uses 1,900 miles of stout copper wire, raised above the ground on poles 35 feet high.

The efficiency of the telephone on a well laid system is so great that it makes very little difference whether the persons talking with one another are 50 or 500 miles apart. There is no reason why a Cape-to-Cairo telephone should not put the two extremities of Africa in clear vocal communication. We may even live to see the day when a London business man will be able to talk with his agent in Sydney, Melbourne, or Wellington.

A step towards this last achievement has been taken by M. Germain, a French electrician, who has patented a telephone which can be used with stronger currents than are possible in ordinary telephones; thereby, of course, increasing the range of speech on submarine cables.

The telephone that we generally use has a transmitter which permits but a small portion of the battery power to pass into the wires, owing to the resistance of the carbon diaphragm. The weakness of the current is to a great extent compensated by the exceedingly delicate nature of the receiver.

M. Germain has reversed the conditions with a transmitter that allows a very high percentage of the current to flow into the wires, and a comparatively insensitive receiver. The result is a 'loud-speaking telephone', not a novelty, for Edison invented one as long ago as 1877, which is capable of reproducing speech in a wonderfully powerful fashion.

M. Germain, with the help of special tubular receivers, has actually sent messages through a line having the same resistance as that of the London-Paris line, so audibly that the words could be heard fifteen yards from the receiver in the open air!

———————————

Chapter 4

Wireless Telephony

In days when wireless telegraphy is occupying such a great deal of the world's attention, it is not likely to cause much astonishment in the reader to learn that wireless transmission of speech over considerable distances is an accomplished fact. We have already mentioned (see 'Wireless Telegraphy') that by means of parallel systems of wires Sir William Preece bridged a large air-gap, and induced in the one sounds imparted to the other.

Since then two other methods have been introduced; and as a preface to the mention of the first we may say a few words about Graham Bell's Photophone.

The Photophone

In this instrument light is made to do the work of a metal connection between speaker and listener. Professor Bell, in arranging the Photophone, used a mouthpiece as in his electric telephone, but instead of a diaphragm working in front of a magnet to set up electric impulses along a wire he employed a mirror of very thin glass, silvered on one side. The effect of sound on this mirror was to cause rapid alterations of its shape from concave to convex, and consequent variations of its reflecting power. A strong beam of light was concentrated on the centre of the mirror through a lens, and reflected by the mirror at an angle through another lens in the direction of the receiving instrument. The receiver consisted of a parabolic reflector to catch the rays and focus them on a selenium cell connected by an electric circuit with an ordinary telephone earpiece.

On delivering a message into the mouthpiece the speaker would, by agitating the mirror, send a succession of light waves of varying intensity towards the distant selenium cell. Selenium has the peculiar property of offering less resistance to electrical currents when light is thrown upon it than when it is in darkness: and the more intense is the light the less is the obstruction it affords. The light-waves from the mirror, therefore, constantly alter its capacity as a conductor, allowing currents to pass through the telephone with varying power.

In this way Professor Bell bridged 800 yards of space; over which he sent, besides articulate words, musical notes, using for the latter purpose a revolving perforated disc to interrupt a constant beam of light a certain number of times per second. As the speed of the disc increased the rate of the light-flashes increased also, and produced in the selenium cell the same number of passages to the electric current, converted into a musical note by the receiver. So that by means of mechanical apparatus a 'playful sunbeam' could literally be compelled to play a tune.

From the Photophone we pass to another method of sound transmission by light, with which is connected the name of Mr. Hammond V. Hayes of Boston, Massachusetts. It is embodied in the Radiophone, or the Ray-speaker, for it makes strong rays of light carry the human voice.

The Radiophone or Ray-speaker

Luminous bodies give off heat. As the light increases, so as a general rule does the heat also. At present we are unable to create strong light without having recourse to heat to help us, since we do not know how to cause other vibrations of sufficient rapidity to yield the sensation of light. But we can produce heat directly, and heat will set atoms in motion, and the ether too, giving us light, but taking as reward a great deal of the energy exerted. Now, the electric arc of a searchlight produces a large amount of light and heat. The light is felt by the eye at a distance of many miles, but the body is not sensitive enough to be aware of the heat emanating from the same source. Mr. Hayes has, however, found the heat accompanying a searchlight beam quite sufficient to affect a mechanical 'nerve' in a far-away telephone receiver.

The transmitting apparatus is a searchlight, through the back of which run four pairs of wires connected with a telephone mouthpiece after passing through a switch and resistance-box or regulator. The receiver is a concave mirror, in the focus of which is a tapering glass bulb, half filled with carbonised filament very sensitive to heat. The tapering end of the bulb projects through the back of the mirror into an ear tube.

If a message is to be transmitted the would-be speaker turns his searchlight in the direction of the person with whom he wishes to converse, and makes the proper signals. On seeing them the other presents his mirror to the beam and listens.

The speaker's voice takes control of the searchlight beam. The louder the sound the more brilliantly glows the electric arc; the stronger becomes the beam, the greater is the amount of heat passed on to the mirror and gathered on the sensitive bulb. The filament inside expands. The tapering point communicates the fact to the earpiece.

This operation being repeated many times a second the earpiece fills with sound, in which all the modulations of the far-distant voice are easily distinguishable.

Two sets of the apparatus above described are necessary for a conversation, the functions of the searchlight and the bulb not being reversible. But inasmuch as all large steamers carry searchlights the necessary installation may be completed at a small expense. Mr. Hayes' invention promises to be a rival to wireless telegraphy over comparatively short distances. It can be relied upon in all weathers, and is a fast method of communication. Like the photophone it illustrates the inter-relationship of the phenomena of Sound, Light, and Heat, and the readiness with which they may be combined to attain an end.

Collins' Earth Transmitter

Next we turn from air to earth, and to the consideration of the work of Mr. A. F. Collins of Philadelphia. This electrician merely makes use of the currents flowing in all directions through the earth, and those excited by an electric battery connected with earth. The outfit requisite for sending wireless spoken messages consists of a couple of convenient stands, as many storage batteries, sets of coils, and receiving and transmitting instruments.

The action of the transmitter is to send from the battery a series of currents through the coils, which transmit them, greatly intensified, to the earth by means of a wire connected with a buried wire screen. The electric disturbances set up in the earth travel in all directions, and strike a similar screen buried beneath the receiving instrument, where the currents affect the delicate diaphragm of the telephone earpiece.

The system is, in fact, upon all fours with Mr. Marconi's, the distinguishing feature being that the ether of the atmosphere is used in the latter case, that of the earth in the former. The intensity coils are common to both; the buried screens are the counterpart of the aerial kites or balloons; the telephone transmitter corresponds to the telegraphic transmitting key; the earpiece to the coherer and relay. No doubt in time Mr. Collins will 'tune' his instruments, so obtaining below ground the same sympathetic electric vibrations which Mr. Marconi, Professor Lodge, or others have employed to clothe their aerial messages in secrecy.

Chapter 5
The Phonograph

Even if Thomas Edison had not done wonders with electric lighting, telephones, electric torpedoes, new processes for separating iron from its ore, telegraphy, animated photography, and other things too numerous to mention, he would still have made for himself an enduring name as the inventor of the Phonograph. He has fitly been called the 'Wizard of the West' from his genius for conjuring up out of what would appear to the multitude most unpromising materials startling scientific marvels, among which none is more truly wizard-like than the instrument that is as receptive of sound as the human ear, and of illimitable reproducing power. By virtue of its elfishly human characteristic, articulate speech, it occupies, and always will occupy, a very high position as a mechanical wonder.

When listening to a telephone we are aware of the fact that the sounds are immediate reproductions of a living person's voice, speaking at the moment and at a definite distance from us; but the phonographic utterances are those of a voice perhaps stilled for ever, and the difference adds romance to the speaking machine.

The Phonograph was born in 1876. As we may imagine, its appearance created a stir. A contributor to the Times wrote in 1877:

'Not many weeks have passed since we were startled by the announcement that we could converse audibly with each other, although hundreds of miles apart, by means of so many miles of wire with a little electric magnet at each end.

Another wonder is now promised us; an invention purely mechanical in its nature, by means of which words spoken by the human voice can be, so to speak, stored up and reproduced at will over and over again hundreds, it may be thousands, of times. What will be thought of a piece of mechanism by means of which a message of any length can be spoken on to a plate of metal; that plate sent by post to any part of the world and the message absolutely respoken in the very voice of the sender, purely by mechanical agency? What, too, shall be said of a mere machine, by means of which the old familiar voice of one who is no longer with us on earth can be heard speaking to us in the very tones and measure to which our ears were once accustomed?'

The Phonautograph

The first Edison machine was the climax of research in the realm of sound. As long ago as 1856 a Mr. Leo Scott made an instrument which received the formidable name of Phonautograph, on account of its capacity to register mechanically the vibrations set up in the atmosphere by the human voice or by musical instruments. A large metal cone like the mouth of an ear-trumpet had stretched across its smaller end a membrane, to which was attached a very delicate tracing-point working on the surface of a revolving cylinder covered with blackened paper. Any sound entering the trumpet agitated the membrane, which in turn moved the stylus and produced a line on the cylinder corresponding to the vibration. Scott's apparatus could only record. It was, so to speak, the first half of the phonograph. Edison, twenty years later, added the active half. His machine, as briefly described in the Times, was simple; so very simple that many scientists must have wondered how they failed to invent it themselves.

A metal cylinder grooved with a continuous square section thread of many turns to the inch was mounted horizontally on a long axle cut at one end with a screw-thread of the same 'pitch' as that on the cylinder. The axle, working in upright supports, and furnished with a heavy fly-wheel to render the rate of revolution fairly uniform, was turned by a handle. Over the grooved cylinder was stretched a thin sheet of tinfoil, and on this rested lightly a steel tracing point, mounted at the end of a spring and separated from a vibrating diaphragm by a small pad of rubber tubing. A large mouthpiece to concentrate sound on to the diaphragm completed the apparatus.

To make a record with this machine the cylinder was moved along until the tracing-point touched one extremity of the foil. The person speaking into the mouthpiece turned the handle to bring a fresh surface of foil continuously under the point, which, owing to the thread on the axle and the groove on the cylinder being of the same pitch, was always over the groove, and burnished the foil down into it to a greater or less depth according to the strength of the impulses received from the diaphragm.

The record being finished, the point was lifted off the foil, the cylinder turned back to its original position, and the point allowed to run again over the depressions it had made in the metal sheet. The latter now became the active part, imparting to the air by means of the diaphragm vibrations similar in duration and quality to those that affected it when the record was being made.

It is interesting to notice that the phonograph principle was originally employed by Edison as a telephone 'relay'. His attention had been drawn to the telephone recently produced by Graham Bell, and to the evil effects of current leakage in long lines. He saw that the amount of current wasted increased out of proportion to the length of the lines-even more than in the proportion of the squares of their lengths; and he hoped that a great saving of current would be effected if a long line were divided into sections and the sound vibrations were passed from one to the other by mechanical means. He used as the connecting link between two sections a strip of moistened paper, which a needle, attached to a receiver, indented with minute depressions, that handed on the message to another telephone. The phonograph proper, as a recording machine, was an after-thought.

Edison's first apparatus, besides being heavy and clumsy, had in practice faults which made it fall short of the description given in the Times. Its tone was harsh. The records, so far from enduring a thousand repetitions, were worn out by a dozen. To these defects must be added a considerable difficulty in adjusting a record made on one machine to the cylinder of another machine.

The Graphophone

Edison, being busy with his telephone and electric lamp work, put aside the phonograph for a time. Graham Bell, his brother, Chichester Bell, and Charles Sumner Tainter, developed and improved his crude ideas. They introduced the Graphophone, using easily removable cylinder records. For the tinfoil was substituted a thin coating of a special wax preparation on light paper cylinders. Clockwork-driven motors replaced the hand motion, and the new machines were altogether more handy and effective. As soon as he had time Edison again entered the field. He conceived the solid wax cylinder, and patented a small shaving apparatus by means of which a record could be pared away and a fresh surface be presented for a new record.

The phonograph or graphophone of to-day is a familiar enough sight; but inasmuch as our readers may be less intimately acquainted with its construction and action than with its effects, a few words will now be added about its most striking features.

In the first place, the record remains stationary while the trumpet, diaphragm and stylus pass over it. The reverse was the case with the tinfoil instrument.

The record is cut by means of a tiny sapphire point having a circular concave end very sharp at the edges, to gouge minute depressions into the wax. The point is agitated by a delicate combination of weights and levers connecting it with a diaphragm of French glass 1 inch thick. The reproducing point is a sapphire ball of a diameter equal to that of the gouge. It passes over the depressions, falling into them in turn and communicating its movements to a diaphragm, and so tenderly does it treat the records that a hundred repetitions do not inflict noticeable damage.

It is a curious instance of the manner in which man unconsciously copies nature that the parts of the reproducing attachment of a phonograph contains parts corresponding in function exactly to those bones of the ear known as the Hammer, Anvil, and Stirrup.

To understand the inner working of the phonograph the reader must be acquainted with the theory of sound. All sound is the result of impulses transmitted by a moving body usually reaching the ear through the medium of the air. The quantity of the sound, or loudness, depends on the violence of the impulse; the tone, or note, on the number of impulses in a given time (usually fixed as one second); and the quality, or timbre, as musicians say, on the existence of minor vibrations within the main ones.

If we were to examine the surface of a phonograph record (or phonogram) under a powerful magnifying glass we should see a series of scoops cut by the gouge in the wax, some longer and deeper than others, long and short, deep and shallow, alternating and recurring in regular groups. The depth, length, and grouping of the cuts decides the nature of the resultant note when the reproducing sapphire point passes over the record, at a rate of about ten inches a second.

The study of a tracing made on properly prepared paper by a point agitated by a diaphragm would enable us to understand easily the cause of that mysterious variation in timbre which betrays at once what kind of instrument has emitted a note of known pitch. For instance, let us take middle C, which is the result of a certain number of atmospheric blows per second on the drum of the ear. The same note may come from a piano, a violin, a banjo, a man's larynx, an organ, or a cornet; but we at once detect its source. It is scarcely imaginable that a piano and a cornet should be mistaken for one another. Now, if the tracing instrument had been at work while the notes were made successively it would have recorded a wavy line, each wave of exactly the same length as its fellows, but varying in its outline according to the character of the note's origin.

We should notice that the waves were themselves wavy in section, being jagged like the teeth of a saw, and that the small secondary waves differed in size.

The minor waves are the harmonics of the main note. Some musical instruments are richer in these harmonics than others. The fact that these delicate variations are recorded as minute indentations in the wax and reproduced is a striking proof of the phonograph's mechanical perfection.

Furthermore, the phonograph registers not only these composite notes, but also chords or simultaneous combinations of notes, each of which may proceed from a different instrument. In its action it here resembles a man who by constant practice is able to add up the pounds, shillings, and pence columns in his ledger at the same time, one wave system overlapping and blending with another.

The phonograph is not equally sympathetic with all classes of sounds. Banjo duets make good records, but the guitar gives a poor result. Similarly, the cornet is peculiarly effective, but the bass drum disappointing. The deep chest notes of a man come from the trumpet with startling truth, but the top notes on which the soprano prides herself are often sadly 'tinny'. The phonograph, therefore, even in its most perfect form is not the equal of the exquisitely sensitive human ear; and this may partially be accounted for by the fact that the diaphragm in both recorder and reproducer has its own fundamental note which is not in harmony with all other notes, whereas the ear, like the eye, adapts itself to any vibration.

Yet the phonograph has an almost limitless repertoire. It can justly be claimed for it that it is many musical instruments rolled into one. It will reproduce clearly and faithfully an orchestra, an instrumental soloist, the words of a singer, a stump orator, or a stage favourite. Consequently we find it everywhere at entertainments, in the drawing-room, and even tempting us at the railway station or other places of public resort to part with our superfluous pence. At the London Hippodrome it discourses to audiences of several thousand persons, and in the nursery it delights the possessors of ingeniously constructed dolls which, on a button being pressed and concealed machinery being brought into action, repeat some well-known childish melody.

It must not be supposed that the phonograph is nothing more than a superior kind of scientific toy. More serious duties than those of mere entertainment have been found for it.

At the last Presidential Election in the States the phonograph was often called upon to harangue large meetings in the interests of the rival candidates, who were perhaps at the time wearing out their voices hundreds of miles away with the same words.

Since the pronunciation of a foreign language is acquired by constant imitation of sounds, the phonograph, instructed by an expert, has been used to repeat words and phrases to a class of students until the difficulties they contain have been thoroughly mastered. The sight of such a class hanging on the lips, or more properly the trumpet, of a phonograph gifted with the true Parisian accent may be common enough in the future.

As a mechanical secretary and substitute for the shorthand writer the phonograph has certainly passed the experimental stage. Its daily use by some of the largest business establishments in the world testify to its value in commercial life. Many firms, especially American, have invested heavily in establishing phonograph establishments to save labour and final expense. The manager, on arriving at his office in the morning, reads his letters, and as the contents of each is mastered, dictates an answer to a phonograph cylinder which is presently removed to the typewriting room, where an assistant, placing it upon her phonograph and fixing the tubes to her ears, types what is required. It is interesting to learn that at Ottawa, the seat of the Canadian Government, phonographs are used for reporting the parliamentary proceedings and debates.

There is therefore a prospect that, though the talking-machine may lose its novelty as an entertainer, its practical usefulness will be largely increased. And while considering the future of the instrument, the thought suggests itself whether we shall be taking full advantage of Mr. Edison's notable invention if we neglect to make records of all kinds of intelligible sounds which have more than a passing interest. If the records were made in an imperishable substance they might remain effective for centuries, due care being taken of them in special depositories owned by the nation. To understand what their value would be to future generations we have only to imagine ourselves listening to the long-stilled thunder of Earl Chatham, to the golden eloquence of Burke, or the passionate declamations of Mrs. Siddons. And in the narrower circle of family interests how valuable a part of family heirlooms would be the phonograms containing a vocal message to posterity from Grandfather this, or Great-aunt that, whose portraits in the drawing-room album do little more than call attention to the changes in dress since the time when their subjects faced the camera!

Record-Making and Manufacture

Phonographic records are of two shapes, the cylindrical and the flat, the latter cut with a volute groove continuously diminishing in diameter from the circumference to the centre. Flat records are used in the Gramophone; a reproducing machine only. Their manufacture is effected by first of all making a record on a sheet of zinc coated with a very thin film of wax, from which the sharp steel point moved by the recording diaphragm removes small portions, baring the zinc underneath.

The plate is then flooded with an acid solution, which eats into the bared patches, but does not affect the parts still covered with wax. The etching complete, the wax is removed entirely, and a cast or electrotype negative record made from the zinc plate. The indentations of the original are in this represented by excrescences of like size; and when the negative block is pressed hard down on to a properly prepared disc of vulcanite or celluloid, the latter is indented in a manner that reproduces exactly the tones received on the 'master' record.

Cylindrical records are made in two ways, by moulding or by copying. The second process is extremely simple. The 'master' cylinder is placed on a machine which also rotates a blank cylinder at a short distance from and parallel to the first. Over the 'master' record passes a reproducing point, which is connected by delicate levers to a cutting point resting on the 'blank', so that every movement of the one produces a corresponding movement of the other.

This method, though accurate in its results, is comparatively slow. The moulding process is therefore becoming the more general of the two. Edison has recently introduced a most beautiful process for obtaining negative moulds from wax positives. Owing to its shape, a zinc cylinder could not be treated like a flat disc, as, the negative made, it could not be detached without cutting. Edison, therefore, with characteristic perseverance, sought a way of electrotyping the wax, which, being a non-conductor of electricity, would not receive a deposit of metal. The problem was how to deposit on it.

Any one who has seen a Crookes' tube such as is used for X-ray work may have noticed on the glass a black deposit which arises from the flinging off from the negative pole of minute particles of platinum. Edison took advantage of this repellent action; and by enclosing his wax records in a vacuum between two gold poles was able to coat them with an infinitesimally thin skin of pure gold, on which silver or nickel could be easily deposited. The deposit being sufficiently thick the wax was melted out and the surface of the electrotype carefully cleaned.

To make castings it was necessary only to pour in wax, which on cooling would shrink sufficiently to be withdrawn. The delicacy of the process may be deduced from the fact that some of the sibilants, or hissing sounds of the voice, are computed to be represented by depressions less than a millionth of an inch in depth, and yet they are most distinctly reproduced! Cylinder records are made in two sizes, 2.5 and 5 inches in diameter respectively. The larger size gives the most satisfactory renderings, as the indentations are on a larger scale and therefore less worn by the reproducing point. One hundred turns to the inch is the standard pitch of the thread; but in some records the number is doubled.

Phonographs, Graphophones, and Gramophones are manufactured almost entirely in America, where large factories, equipped with most perfect plant and tools, work day and night to cope with the orders that flow in freely from all sides. One factory alone turns out a thousand machines a day, ranging in value from a few shillings to forty pounds each. Records are made in England on a large scale; and now that the Edison-Bell firm has introduced the unbreakable celluloid form their price will decrease. By means of the Edison electrotyping process a customer can change his record without changing his cylinder. He takes the cylinder to the factory, where it is heated, placed in the mould, and subjected to great pressure which drives the soft celluloid into the mould depressions; and behold! in a few moments 'Auld Lang Syne' has become 'Home, Sweet Home', or whatever air is desired. Thus altering records is very little more difficult than getting a fresh book at the circulating library

The Photographophone

This instrument is a phonograph working entirely by means of light and electricity. The flame of an electric lamp is brought under the influence of sound vibrations which cause its brilliancy to vary at every alteration of pitch or quality.

The light of the flame is concentrated through a lens on to a travelling photographic sensitive film, which, on development in the ordinary way, is found to be covered with dark and bright stripes proportionate in tone to the strength of the light at different moments. The film is then passed between a lamp and a selenium plate connected with an electric circuit and a telephone. The resistance of the selenium to the current varies according to the power of the light thrown upon it. When a dark portion of the film intercepts the light of the lamp the selenium plate offers high resistance; when the light finds its way through a clear part of the film the resistance weakens. Thus the telephone is submitted to a series of changes affecting the 'receiver'. As in the making of the record speech-vibrations affect light, and the light affects a sensitive film; so in its reproduction the film affects a sensitive selenium plate, giving back to a telephone exactly what it received from the sound vibrations.

One great advantage of Mr. Ruhmer's method is that from a single film any number of records can be printed by photography; another, that, as with the Telegraphone, the same film passed before a series of lamps successively is able to operate a corresponding number of telephones.

The inventor is not content with his success. He hopes to record not merely sounds but even pictures by means of light and a selenium plate.

Chapter 6

The Telephonograph & Others

Having dealt with the phonograph and the telephone separately, we may briefly consider one or two ingenious combinations of the two instruments. The word Telephonograph signifies an apparatus for recording sounds sent from a distance. It takes the place of the human listener at the telephone receiver.

Let us suppose that a Reading subscriber wishes to converse along the wires with a friend in London, but that on ringing up his number he discovers that the friend is absent from his home or office. He is left with the alternative of either waiting till his friend returns, which may cause a serious loss of time, or of dictating his message, a slow and laborious process. This with the ordinary telephonic apparatus. But if the London friend be the possessor of a Telephonograph, the person answering the call-bell can, if desired to do so, switch the wires into connection with it and start the machinery; and in a very short time the message will be stored up for reproduction when the absent friend returns.

The Telephonograph is the invention of Mr. J. E. O. Kumberg. The message is spoken into the telephone transmitter in the ordinary way, and the vibrations set up by the voice are caused to act upon a recording stylus by the impact of the sound waves at the further end of the wires. In this manner a phonogram is produced on the wax cylinder in the house or office of the person addressed, and it may be read off at leisure. A very sensitive transmitter is employed, and if desired the apparatus can be so arranged that by means of a double-channel tube the words spoken are simultaneously conveyed to the telephone and to an ordinary phonograph, which insures that a record shall be kept of any message sent.

The Telegraphone

The Telegraphone, produced by Mr. Valdemar Poulsen, performs the same functions as the telephonograph, but differs from it in being entirely electrical. It contains no waxen cylinder, no cutting-point; their places are taken respectively by a steel wire wound on a cylindrical drum (each turn carefully insulated from its neighbours) and by a very small electro-magnet, which has two delicate points that pass along the wire, one on either side, resting lightly upon it.

As the drum rotates, the whole of the wire passes gradually between the two points, into which a series of electric shocks is sent by the action of the speaker's voice at the further end of the wires. The shocks magnetise the portion of steel wire which acts as a temporary bridge between the two points. At the close of three and a half minutes the magnet has worked from one end of the wire coil to the other; it is then automatically lifted and carried back to the starting-point in readiness for reproduction of the sounds. This is accomplished by disconnecting the telegraphone from the telephone wires and switching it on to an ordinary telephonic earpiece or receiver. As soon as the cylinder commences to revolve a second time, the magnet is influenced by the series of magnetic 'fields' in the wires, and as often as it touches a magnetised spot imparts an impulse to the diaphragm of the receiver, which vibrates at the rate and with the same force as the vibrations originally set up in the distant transmitter. The result is a clear and accurate reproduction of the message, even though hours and even days may have elapsed since its arrival.

As the magnetic effects on the wire coil retain their power for a considerable period, the message may be reproduced many times.

As soon as the wire-covered drum is required for fresh impressions, the old one is wiped out by passing a permanent magnet along the wire to neutralise the magnetism of the last message.

Mr. Poulsen has made an instrument of a different type to be employed for the reception of an unusually lengthy communication. Instead of a wire coil on a cylinder, a ribbon of very thin flat steel spring is wound from one reel on to another across the poles of two electro-magnets, which touch the lower side only of the strip. The first magnet is traversed by a continuous current to efface the previous record; the second magnetises the strip in obedience to impulses from the telephone wires. The message complete, the strip is run back, and the magnets connected with receivers, which give out loud and intelligent speech as the strip again traverses them. The Poulsen machine makes the transmission of the same message simultaneously through several telephones an easy matter, as the strip can be passed over a series of electro-magnets each connected with a telephone.

The Telautograph

It is a curious experience to watch for the first time the movements of a tiny telautograph pen as it works behind a glass window in a japanned case. The pen, though connected only with two delicate wires, appears instinct with human reason. It writes in a flowing hand, just as a man writes. At the end of a word it crosses the t's and dots the i's. At the end of a line it dips itself in an ink pot. It punctuates its sentences correctly. It illustrates its words with sketches. It uses shorthand as readily as longhand. It can form letters of all shapes and sizes.

And yet there is no visible reason why it should do what it does. The japanned case hides the guiding agency, whatever it may be. Our ears cannot detect any mechanical motion. The writing seems at first sight as mysterious as that which appeared on the wall to warn King Belshazzar.

In reality it is the outcome of a vast amount of patience and mechanical ingenuity culminating in a wonderful instrument called the telautograph. The telautograph is so named because by its aid we can send our autographs, i.e, our own particular handwriting, electrically over an indefinite length of wire, as easily as a telegraph clerk transmits messages in the Morse alphabet. Whatever the human hand does on one telautograph at one end of the wires, that will be reproduced by a similar machine at the other end, though the latter be hundreds of miles away.

The working speed of the telautograph is that of the writer. If shorthand be employed, messages can be transmitted at the rate of over 100 words per minute. As regards the range of transmission, successful tests have been made by the postal authorities between Paris and London, and also between Paris and Lyons. In the latter case the messages were sent from Paris to Lyons and back directly to Paris, the lines being connected at Lyons, to give a total distance of over 650 miles. There is no reason why much greater length of line should not be employed.

The telautograph in its earlier and imperfect form was the work of Professor Elisha Gray, who invented the telephone almost simultaneously with Professor Graham Bell. His telautograph worked on what is known as the step-by-step principle, and was defective in that its speed was very limited. If the operator wrote too fast the receiving pen lagged behind the transmitting pencil, and confusion resulted. Accordingly this method, though ingenious, was abandoned, and Mr. Ritchie in his experiments looked about for some preferable system, which should be simpler and at the same time much speedier in its action. After four years of hard work he has brought the rheostat system, explained above, to a pitch of perfection which will be at once appreciated by any one who has seen the writing done by the instrument.

The advantages of the Telautograph over the ordinary telegraphy may be briefly summed up as follows:

Anybody who can write can use it; the need of skilled operators is abolished.

A record is automatically kept of every message sent.

The person to whom the message is sent need not be present at the receiver. He will find the message written out on his return.

The instrument is silent and so insures secrecy. An ordinary telegraph may be read by sound; but not the telautograph.

It is impossible to tap the wires unless, as is most unlikely, the intercepting party has an instrument in exact accord with the transmitter.

It can be used on the same wires as the ordinary telephone, and since a telephone is combined with it, the subscriber has a double means of communication. For some items of business the telephone may be used as preferable ; but in certain cases, the telautograph. A telephone message may be heard by other subscribers; it is impossible to prove the authenticity of such a message unless witnesses have been present at the transmitting end; and the message itself may be misunderstood by reason of bad articulation. But the telautograph preserves secrecy while preventing any misunderstanding. Anything written by it is for all practical purposes as valid as a letter.

We must not forget its extreme usefulness for transmitting sketches. A very simple diagram often explains a thing better than pages of letter-press. The telautograph may help in the detection of criminals, a pictorial presentment of whom can by its means be despatched all over the country in a very short time. And in warfare an instrument flashing back from the advance-guard plans of the country and of the enemy's positions might on occasion prove of the greatest importance.

Section Two

Transport

Chapter 7

Horseless Carriages

A body of enterprising Manchester merchants, in the year 1754, put on the road a 'flying coach', which, according to their special advertisement, would, 'however incredible it may appear, actually, barring accidents, arrive in London in four and a half days after leaving Manchester'. According to the Lord Chancellor of the time such swift travelling was considered dangerous as well as wonderful, the condition of the roads might well make it so, and also injurious to health. 'I was gravely advised', he says, 'to stay a day in York on my journey between Edinburgh and London, as several passengers who had gone through without stopping had died of apoplexy from the rapidity of the motion'.

As the coach took a fortnight to pass from the Scotch to the English capital, at an average pace of between three and four miles an hour, it is probable that the Chancellor's advisers would be very seriously indisposed by the mere sight of a motor-car whirling along in its attendant cloud of dust, could they be resuscitated for the purpose. And we, on the other hand, should prefer to get out and walk to 'flying' at the safe speed of their mail coaches.

The improvement of highroads, and road-making generally, accelerated the rate of posting. In the first quarter of the nineteenth century an average of ten or even twelve miles an hour was maintained on the Bath Road. But that pace was considered inadequate when the era of the 'iron horse' commenced, and the decay of stage-driving followed hard upon the growth of railways. What should have been the natural successor of the stage-coach was driven from the road by ill-advised legislation, which gave the railroads a monopoly of swift transport, which has but lately been removed.

The history of the steam-coach, steam-carriage, automobile, motor-car to give it its successive names, is in a manner unique, showing as it does, instead of steady development of a practical means of locomotion, a sudden and decisive check to an invention worthy of far better treatment than it received. The compiler of even a short survey of the automobile's career is obliged to divide his account into two main portions, linked together by a few solitary engineering achievements.

The First Period (1800 – 1836)

This will, without any desire to arrogate for England more than her due or to belittle the efforts of any other nations, be termed the English Period, since in it England took the lead, and produced by far the greatest number of steam-carriages. The second (1870 to the present day) may, with equal justice, be styled the Continental Period, as witnessing the great developments made in automobilism by French, German, Belgian, and American engineers: England, for reasons that will be presently noticed, being until quite recently too heavily handicapped to take a part in the advance.

Historically, it is impossible to discover who made the first self-moving carriage. In the sixteenth century one Johann Haustach, a Nuremberg watch-maker, produced a vehicle that derived its motive power from coiled springs, and was in fact a large edition of our modern clockwork toys. About the same time the Dutch, and among them especially one Simon Stevin, fitted carriages with sails, and there are records of a steam-carriage as early as the same century.

But the first practical, and at least semi-successful, automobile driven by internal force was undoubtedly that of a Frenchman, Nicholas Joseph Cugnot, who justly merits the title of father of automobilism. His machine, which is today one of the most treasured exhibits in the Paris Museum of Arts and Crafts, consisted of a large carriage, having in front a pivoted platform bearing the machinery, and resting on a solid wheel, which propelled as well as steered the vehicle. The boiler, of stout riveted copper plates, had below it an enclosed furnace, from which the flames passed upwards through the water through a funnel. A couple of cylinders, provided with a simple reversing gear, worked a ratchet that communicated motion to the driving-wheel. This carriage did not travel beyond a very slow walking pace, and Cugnot therefore added certain improvements, after which (1770) it reached the still very moderate speed of four miles an hour, and distinguished itself by charging and knocking down a wall, a feat that is said, to have for a time deterred engineers from developing a seemingly dangerous mode of progression.

Ten years later Dallery built a steam car, and ran it in the streets of Amiens; we are not told with what success; and before any further advance had been made with the automobile the French Revolution put a stop to all inventions of a peaceful character among our neighbours.

In England, however, steam had already been recognised as the coming power. Richard Trevethick, afterwards to become famous as a railroad engineer, built a steam motor in 1802, and actually drove it from Cambourne to Plymouth, a distance of ninety miles. But instead of following up this success, he forsook steam-carriages for the construction of locomotives, leaving his idea to be expanded by other men, who were convinced that a vehicle which could be driven over existing roads was preferable to one that was helpless when separated from smooth metal rails. Between the years 1800 and 1836 many steam vehicles for road traffic appeared from time to time, some, such as David Gorton's (propelled by metal legs pressing upon the ground), strangely unpractical, but the majority showing a steady improvement in mechanical design.

As it will be impossible, without writing a small book, to name all the English constructors of this period, we must rest content with the mention of the leading pioneers of the new locomotion.

Sir Goldsworthy Gurney, an eminent chemist, did for mechanical road propulsion what George Stephenson was doing for railway development. He boldly spent large sums on experimental vehicles, which took the form of six-wheeled coaches. The earliest of these were fitted with legs as well as driving-wheels, since he thought that in difficult country wheels alone would not have sufficient grip. (A similar fallacy was responsible for the cogged wheels on the first railways.) But in the later types legs were abandoned as unnecessary. His coaches easily climbed the steepest hills round London, including Highgate Hill, though a thoughtful mathematician had proved by calculations that a steam carriage, so far from mounting a gradient, could not, without violating all natural laws, so much as move itself on the level!

Having satisfied himself of their power, Gurney took his coaches further afield. In 1829 was published the first account of a motor trip made by him and three companions through Reading, Devizes, and Melksham. The pace was, we read, at first only about six miles an hour, including stoppages. They drove very carefully to avoid injury to the persons or feelings of the country folk; but at Melksham, where a fair was in progress, they had to face a shower of stones, hurled by a crowd of roughs at the instigation of some coaching postilions, who feared losing their livelihood if the new method of locomotion became general. Two of the tourists were severely hurt, and Gurney was obliged to take shelter in a brewery, where constables guarded his coach. On the return journey the party timed their movements so as to pass through Melksham while the inhabitants were all safely in bed.

The coach ran most satisfactorily, improving every mile. 'Our pace was so rapid', wrote one of the company, 'that the horses of the mail-cart which accompanied us were hard put to it to keep up with us. At the foot of Devizes Hill we met a coach and another vehicle, which stopped to see us mount this hill, an extremely steep one. We ascended it at a rapid rate. The coach and passengers, delighted at this unexpected sight, honoured us with shouts of applause'.

In 1830 Messrs. Ogle and Summers completely beat the road record on a vehicle fitted with a tubular boiler. This car, put through its trials before a Special Commission of the House of Commons, attained the astonishing speed of 35 miles an hour on the level, and mounted a hill near Southampton at 21 miles an hour. It worked at a boiler pressure of 250 pounds to the square inch, and though not hung on springs, ran 800 miles without a breakdown. This performance appears all the more extraordinary when we remember the roads of that day were not generally as good as they are now, and that in the previous year Stephenson's 'Rocket', running on rails, had not reached a higher velocity.

The report of the Parliamentary Commission on horseless carriages was most favourable. It urged that the steam-driven car was swifter and lighter than the mail-coaches; better able to climb and descend hills; safer; more economical; and less injurious to the roads; and, in conclusion, that the heavy charges levied at the toll-gates (often twenty times those on horse vehicles) were nothing short of iniquitous.

As a result of this report, motor services, inaugurated by Walter Hancock, Braithwayte, and others, commenced between Paddington and the Bank, London and Greenwich, London and Windsor, London and Stratford. Already, in 1829, Sir Charles Dance had a steam-coach running between Cheltenham and Gloucester. In four months it ran 3,500 miles and carried 3,000 passengers, traversing the nine miles in three-quarters of an hour; although narrow-minded landowners placed ridges of stone eighteen inches deep on the road by way of protest.

The most ambitious service of all was that between London and Birmingham, established in 1833 by Dr. Church. The rolling-stock consisted of a single very much decorated coach.

The success of the road-steamer seemed now assured, when a cloud appeared on the horizon. It had already been too successful. The railway companies were up in arms. They saw plainly that if once the roads were covered with vehicles able to transport the public at low fares quickly from door-to-door on existing thoroughfares, the construction of expensive railroads would be seriously hindered, if not altogether stopped. So, taking advantage of two motor accidents, the companies appealed to Parliament – full of horse-loving squires and manufacturers, who scented profit in the railways – and though scientific opinion ran strongly in favour of the steam-coach, a law was passed in 1836 which rendered the steamers harmless by robbing them of their speed. The fiat went forth that in future every road locomotive should be preceded at a distance of a hundred yards by a man on foot carrying a red flag to warn passengers of its approach.

This law marks the end of the first period of automobilism as far as England is concerned. At one blow it crippled a great industry, deprived the community of a very valuable means of transport, and crushed the energies of many clever inventors who would soon, if we may judge by the rapid advances already made in construction, have brought the steam-carriage to a high pitch of perfection. In the very year in which they were suppressed the steam services had proved their efficiency and safety. Hancock's London service alone traversed 4,200 miles without serious accident, and was so popular that the coaches were generally crowded. It is therefore hard to believe that these vehicles did not supply a public want, or that they were regarded by those who used them as in any way inferior to horse-drawn coaches. Yet ignorant prejudice drove them off the road for sixty years; and to-day it surprises many Englishmen to learn that what is generally considered a novel method of travelling was already fairly well developed in the time of their grandfathers.

The Second Period (1870 onwards)

To follow the further development of the automobile we must cross the Channel once again. French invention had not been idle while Gurney and Hancock were building their coaches. In 1835 M. Dietz established a service between Versailles and Paris, and the same year M. D'Asda carried out some successful trials of his steam 'diligence' under the eyes of Royalty. But we find that for the next thirty-five years the steam-carriage was not much improved, owing to want of capital among its French admirers. No Gurney appeared, ready to spend his thousands in experimenting; also, though the law left road locomotion unrestricted, the railways offered a determined opposition to a possibly dangerous rival. So that, on the whole, road transport by steam fared badly till after the terrible Franco-Prussian war, when inventors again took courage. M. Bollee, of Mans, built in 1873 a car, 'l'Obeissante', which ran from Mans to Paris; and became the subject of allusions in popular songs and plays, while its name was held up as an example to the Paris ladies. Three years later he constructed a steam omnibus to carry fifty persons, and in 1878 exhibited a car that journeyed at the rate of eighteen miles an hour from Paris to Vienna, where it aroused great admiration.

After the year 1800 French engineers divided their attention between the heavy motor omnibus and light vehicles for pleasure parties. In 1884 M. Bouton and Trepardoux, working conjointly with the Comte de Dion, produced a steam-driven tricycle, and in 1887 M. Serpollet followed suit with another, fitted with the peculiar form of steam generator that bears his name. Then came in 1890 a very important innovation, which has made automobilism what it now is. Gottlieb Daimler, a German engineer, introduced the petrol gas-motor. Its comparative lightness and simplicity at once stamped it as the thing for which makers were waiting. Petrol-driven vehicles were soon abroad in considerable numbers and varieties, but they did not attract public attention to any great extent until, in 1894, M. Pierre Giffard, an editor of the Petit Journal, organised a motor race from Paris to Rouen. The proprietors of the paper offered handsome prizes to the successful competitors. There were ten starters, some on steam, others on petrol cars. The race showed that, so far as stability went, Daimler's engine was the equal of the steam cylinder. The next year another race of a more ambitious character was held, the course being from Paris to Bordeaux and back. Subscriptions for prizes flowed in freely. Serpollet, de Dion, and Bollee prepared steam cars that should win back for steam its lost supremacy, while the petrol faction secretly built motors of a strength to relegate steam once and for all to a back place. Electricity, too, made a bid unsuccessfully for the prize in the Jeantaud car, a special train being engaged in advance to distribute charged accumulators over the route. The steamers broke down soon after the start, so that the petrol cars 'walked over' and won a most decisive victory.

The interest roused in the race led the Comte de Dion to found the Automobile Club of France, which drew together all the enthusiastic admirers of the new locomotion. Automobilism now became a sport, a craze. The French, with their fine straight roads, and a not too deeply ingrained love of horseflesh, gladly welcomed the flying car, despite its noisy and malodorous properties.

Orders flowed in so freely that the motor makers could not keep pace with the demand, or promise delivery within eighteen months. Rich men were therefore obliged to pay double prices if they could find anyone willing to sell, a state of things that remains unto this day with certain makes of French cars. Poorer folks contented themselves with De Dion motor tricycles, which showed up so well in the 1896 Paris-Marseilles race; or with the neat little three-wheeled cars of M. Bollee. Motor racing became the topic of the hour. Journals were started for the sole purpose of recording the doings of motorists; and few newspapers of any popularity omitted a special column of motor news. Successive contests on the highroads at increasing speeds attracted increased interest. The black-goggled, fur-clad chauffeur who carried off the prizes found himself a hero.

In short, the hold which automobiles has over our neighbours may be gauged from the fact that in 1901 it was estimated that nearly a thousand motor cars assembled to see the sport on the Longchamps Course (the scene of that ultra-horsey event, the Grand Prix), and the real interest of the meet did not centre round horses of flesh and blood.

The French have not a monopoly of devotion to automobilism. The speedy motor car is too much in accord with the bustling spirit of the age; its delights too easily appreciated to be confined to one country. Allowing France the first place, America, Germany, and Belgium are not far behind in their addiction to the 'sport', and even in Britain, partially freed since 1896 from the red-flag tyranny, thanks to the efforts of Sir David Salomons, there are most visible signs that the era of the horse is beginning its end.

Types of Car

Automobiles may be classified according to the purpose they serve, according to their size and weight, or according to their motive power. We will first review them under the latter head.

Petrol

The petrol motor, suitable alike for large cars of 40 to 60 horse-power and for the small bicycle weighing 70 pounds or so, at present undoubtedly occupies the first place in popular estimation on account of its comparative simplicity which more than compensates certain defects that affect persons off the vehicle more than those on it; smell and noise.

The chief feature of the internal explosion motor is that at one operation it converts fuel directly into energy, by exploding it inside a cylinder. It is herein more economical than steam, which loses power while passing from the boiler to the driving gear.

Petrol cycles and small cars have usually only one cylinder, but large vehicles carry two, three, and sometimes four cylinders. Four and more avoid that bugbear of rotary motion, 'dead points', during which the momentum of the machinery alone is doing work; and for that reason the engines of racing cars are often quadrupled.

For the sake of simplicity we will describe the working of a single cylinder, leaving the reader to imagine it acting alone or in concert with others as he pleases.

In the first place the fuel, petrol, is a very inflammable distillation of petroleum: so ready to ignite that it must be most rigourously guarded from naked lights; so quick to evaporate that the receptacles containing it, if not quite airtight, will soon render it 'stale' and unprofitable for motor driving.

The engine, to mention its most important parts, consists of a single-action cylinder (giving a thrust one way only); a heavy flywheel revolving in an airtight circular case, and connected to the piston by a hinged rod which converts the reciprocating movement of the piston into a rotary movement of the crank-shaft built in with the wheel; inlet and outlet valves; a carburettor for generating petrol gas, and a device to ignite the gas-and-air mixture in the cylinder.

The action of the engine is as follows: as the piston moves outwards in its first stroke it sucks through the inlet valve a quantity of mixed air and gas, the proportions of which are regulated by special taps. The stroke ended, the piston returns, compressing the mixture and rendering it more combustible. Just as the piston commences its second outward stroke an electric spark passed through the mixture mechanically ignites it, and creates an explosion, which drives the piston violently forwards. The second return forces the burnt gas through the exhaust-valve, which is lifted by cog-gear once in every two revolutions of the crank, into the 'silencer:' The cycle of operations is then repeated.

We see that, during three-quarters of the cycle (the suction, compression, and expulsion), the work is performed entirely by the fly-wheel. It follows that a single-cylinder motor, to work at all, must rotate the wheel at a high rate. Once stopped, it can be restarted only by the action of the handle or pedals; a task often so unpleasant and laborious that the driver of a car, when he comes to rest for a short time only, disconnects his motor from the driving-gear and lets it throb away idly beneath him.

The means of igniting the gas in the cylinders may be either a Bunsen burner or an electric spark. Tube ignition is generally considered inferior to electrical because it does not permit 'timing' of the explosion. Large cars are often fitted with both systems, so as to have one in reserve should the other break down.

Electrical ignition is most commonly produced by the aid of an intensity coil, which consists of an inner core of coarse insulated wire, called the primary coil; and an outer, or secondary coil, of very fine wire. A current passes at intervals, timed by a cam on the exhaust-valve gear working a make-and-break contact blade, from an accumulator through the primary coil, exciting by induction a current of much greater intensity in the secondary. The secondary is connected to a 'sparking plug', which screws into the end of the cylinder, and carries two platinum points about 1/32 of an inch apart. The secondary current leaps this little gap in the circuit, and the spark, being intensely hot, fires the compressed gas. Instead of accumulators a small dynamo, driven by the motor, is sometimes used to produce the primary current.

By moving a small lever, known as the 'advancing lever', the driver can control the time of explosion relatively to the compression of the gas, and raise or lower the speed of the motor.

The strokes of the petrol-driven cylinder are very rapid, varying from 1,000 to 3,000 a minute. The heat of very frequent explosions would soon make the cylinder too hot to work were not measures adopted to keep it cool. Small cylinders, such as are carried on motor cycles, are sufficiently cooled by a number of radiating ribs cast in a piece with the cylinder itself; but for large machines a water jacket or tank surrounding the cylinder is a necessity. Water is circulated through the jacket by means of a small centrifugal pump working off the driving gear, and through a coil of pipes fixed in the front of the car to catch the draught of progression. So long as the jacket and tubes are full of water the temperature of the cylinder cannot rise above boiling point.

Motion is transmitted from the motor to the driving-wheels by intermediate gear, which in cycles may be only a leather band or couple of cogs, but in cars is more or less complicated. Under the body of the car, running usually across it, is the countershaft, fitted at each end with a small cog which drives a chain passing also over much larger cogs fixed to the driving-wheels. The countershaft engages with the cylinder mechanism by a 'friction clutch', a couple of circular faces which can be pressed against one another by a lever.

To start his car the driver allows the motor to obtain a considerable momentum, and then, using the friction lever, brings more and more stress on to the countershaft until the friction-clutch overcomes the inertia of the car and produces movement.

Gearing suitable for level stretches would not be sufficiently powerful for hills: the motor would slow and probably stop from want of momentum. A car is therefore fitted with changing gears, which give two or three speeds, the lower for ascents, the higher for the level: and on declines the friction-clutch can be released, allowing the car to 'coast'.

Steam Cars

Though the petrol car has come to the front of late years it still has a powerful rival in the steam car. Inventors have made strenuous efforts to provide steam-engines light enough to be suitable for small pleasure cars. At present the Locomobile (American) and Serpollet (French) systems are increasing their popularity. The Locomobile, the cost of which contrasts favourably with that of even the cheaper petrol cars, has a small multi-tubular boiler wound on the outside with two or three layers of piano wire, to render it safe at high pressures.

As the boiler is placed under the seat it is only fit and proper that it should have a large margin of safety. The fuel, petrol, is passed through a specially designed burner, pierced with hundreds of fine holes arranged in circles round air inlets. The feed-supply to the burner is governed by a spring valve, which cuts off the petrol automatically as soon as the steam in the boiler reaches a certain pressure. The locomobile runs very evenly and smoothly, and with very little noise, a welcome change after the very audible explosion motor.

The Serpollet system is a peculiar method of generating steam. The boiler is merely a long coil of tubing, into which a small jet of water is squirted by a pump at every stroke of the cylinders. The steam is generated and used in a moment, and the speed of the machine is regulated by the amount of water thrown by the pumps. By an ingenious device the fuel supply is controlled in combination with the water supply, so that there may not be any undue waste in the burner.

Electricity

Of electric cars there are many patterns, but at present they are not commercially so practical as the other two types. The great drawbacks to electrically-driven cars are the weight of the accumulators (which often scale nearly as much as all the rest of the vehicle), and the difficulty of getting them recharged when exhausted. We might add to these the rapidity with which the accumulators become worn out, and the consequent expense of renewal. T. A. Edison is reported at work on an accumulator which will surpass all hitherto constructed, having a much longer life, and weighing very much less, power for power.

The longest continuous run ever made with electricity, 187 miles at Chicago, compares badly with the feat of a petrol car which on November 23, 1900, travelled a thousand miles on the Crystal Palace track in 48 hours 24 minutes, without a single stop. Successful attempts have been made by M. Pieper and Jenatsky to combine the petrol and electric systems, by an arrangement which instead of wasting power in the cylinders when less speed is required, throws into action electric dynamos to store up energy, convertible, when needed, into motive power by reversing the dynamo into a motor.

But the simple electric car will not be a universal favourite until either accumulators are so light that a very large store of electricity can be carried without inconvenient addition of weight, or until charging stations are erected all over the country at distances of fifty miles or so apart.

Whether steam will eventually get the upper hand of the petrol engine is at present uncertain. The steam car has the advantage over the gas-engine car in ease of starting, the delicate regulation of power, facility of reversing, absence of vibration, noise and smell, and freedom from complicated gears.

On the other hand the petrol car has no boiler to get out of order or burst, no troublesome gauges requiring constant attention, and there is small difficulty about a supply of fuel. Petrol sufficient to give motive power for hundreds of miles can be carried if need be; and as long as there is petrol on board the car is ready for work at a moment's notice. Judging by the number of the various types of vehicles actually at work we should say that while steam is best for heavy traction, the gas-engine is most often employed on pleasure cars.

Liquid Air

This will also have to be reckoned with as a motive power. At present it is only on its probation; but the writer has good authority for stating that before these words appear in print there will be on the roads a car driven by liquid air, and able to turn off eighty miles in the hour.

Manufacture

As the English were the pioneers of the steam car, so are the Germans and French the chief manufacturers of the petrol car. While the hands of English manufacturers were tied by short-sighted legislation, continental nations were inventing and controlling valuable patents, so that even now our manufacturers are greatly handicapped. Large numbers of petrol cars are imported annually from France, Germany, and Belgium. Steam cars come chiefly from America and France. The former country sent us nearly 2,000 vehicles in 1901. There are signs, however, that English engineers mean to make a determined effort to recover lost ground; and it is satisfactory to learn that in heavy steam vehicles, such as are turned out by Thorneycroft and Co., this country holds the lead. We will hope that in a few years we shall be exporters in turn.

Having glanced at the history and nature of the various types of car, it will be interesting to turn to a consideration of their travelling capacities. As we have seen, a steam omnibus attained, in 1830, a speed of no less than thirty-five miles an hour on what we should call bad roads. It is therefore to be expected that on good modern roads the latest types of car would be able to eclipse the records of seventy years ago. That such has indeed been the case is evident when we examine the performances of cars in races organised as tests of speed. France, with its straight, beautifully-kept, military roads, is the country par excellence for the chauffeur. One has only to glance at the map to see how the main highways conform to Euclid's dictum that a straight line is the shortest distance between any two points, e.g. between Rouen and Dieppe, where a park of artillery, well posted, could rake the road either way for miles.

The growth of speed in the French races is remarkable. In 1894 the winning car ran at a mean velocity of thirteen miles an hour; in 1895, of fifteen. The year 1898 witnessed a great advance to twenty three miles, and the next year to thirty miles. But all these speeds paled before that of the Paris to Bordeaux race of 1901, in which the winner, M. Fournier, traversed the distance of 327.5 miles at a rate of 53.75 miles per hour! The famous Sud-Express, running between the same cities, and considered the fastest long-distance express in the world, was beaten by a full hour. It is interesting to note that in the same races a motor bicycle, a Werner, weighing 80 pounds or less, successfully accomplished the course at an average rate of nearly thirty miles an hour. The motor-car, after waiting seventy years, had had its revenge on the railways.

This was not the only occasion on which an express service showed up badly against its nimble rival of the roads. In June, 1901, the French and German authorities forgot old animosities in a common enthusiasm for the automobile, and organised a race between Paris and Berlin. It was to be a big affair, in which the cars of all nations should fight for the speed championship. Every possible precaution was taken to insure the safety of the competitors and the spectators. Flags of various colours and placards marked out the course, which lay through Rheims, Luxembourg, Coblentz, Frankfurt, Eisenach, Leipsic, and Potsdam to the German capital. About fifty towns and large villages were 'neutralised', that is to say, the competitors had to consume a certain time in traversing them. At the entrance to each neutralised zone a 'control' was established. As soon as a competitor arrived, he must slow down, and a card on which was written the time of his arrival was handed to a 'pilot', who cycled in front of the car to the other 'control' at the farther end of the zone, from which, when the proper time had elapsed, the car was dismissed.

Among other rules were: that no car should be pushed or pulled during the race by any one else than the passengers; that at the end of the day only a certain time should be allowed for cleaning and repairs; and that a limited number of persons, varying with the size of the car, should be permitted to handle it during that period.

A small army of automobile club representatives, besides thousands of police and soldiers, were distributed along the course to restrain the crowds of spectators. It was absolutely imperative that for vehicles propelled at a rate of from 50 to 60 miles an hour a clear path should be kept.

At dawn, on July 27th, 109 racing machines assembled at the Fort de Champigny, outside Paris, in readiness to start for Berlin. Just before half past three, the first competitor received the signal; two minutes later the second; and then at short intervals for three hours the remaining 107, among whom was one lady, Mme. de Gast. At least 20,000 persons were present, even at that early hour, to give the racers a hearty farewell, and demonstrate the interest attaching in France to all things connected with automobilism.

Great excitement prevailed in Paris during the three days of the race. Every few minutes telegrams arrived from posts on the route telling how the competitors fared. The news showed that during the first stage at least a hard fight for the leading place was in progress. The French cracks, Fournier, Charron, De Knyff, Farman, and Girardot pressed hard on Hourgieres. No. 2 at the starting point, Fournier soon secured the lead, and those who remembered his remarkable driving in the Paris Bordeaux race at once selected him as the winner. Aix-la-Chapelle, 283 miles from Paris and the end of the first stage, was reached in 6 hours 28 minutes. Fournier first, De Knyff second by six minutes.

On the 28th the racing became furious. Several accidents occurred. Edge, driving the only English car, wrecked his machine on a culvert, the sharp curve of which flung the car into the air and broke its springs. Another ruined his chances by running over and killing a boy. But Fournier, Antony, De Knyff, and Girardot managed to avoid mishaps for that day, and covered the ground at a tremendous pace. At Dusseldorf Girardot won the lead from Fournier, to lose it again shortly. Antony, driving at a reckless speed, gained ground all day, and arrived a close second at Hanover, the halting-place, after a run averaging, in spite of bad roads and dangerous corners, no less than 54 miles an hour!

The chauffeur in such a race must indeed be a man of iron nerves. Through the great black goggles which shelter his face from the dust-laden hurricane set up by the speed he travels at he must keep a perpetual, piercingly keen watch. Though travelling at express speed, there are no signals to help him; he must be his own signal-man as well as driver. He must mark every loose stone on the road, every inequality, every sudden rise or depression; he must calculate the curves at the corners and judge whether his mechanician, hanging out on the inward side, will enable a car to round a turn without slackening speed. His calculations and decisions must be made in the fraction of a second, for a moment's hesitation might be disaster. His driving must be furious and not reckless; the timid chauffeur will never win, the careless one will probably lose. His head must be cool although the car leaps beneath him like a wild thing, and the wind lashes his face. At least one well-tried driver found the mere mental strain too great to bear, and retired from the contest; and we may be sure that few of the competitors slept much during the nights of the race.

At four o'clock on the 29th Fournier started on the third stage, which witnessed another bout of fast travelling. It was now a struggle between him and Antony for first place. The pace rose at times to eighty miles an hour, a speed at which our fastest expresses seldom travel. Such a speed means huge risks, for stopping, even with the powerful brakes fitted to the large cars, would be a matter of a hundred yards or more. Not far from Hanover Antony met with an accident, Girardot now held second place; and Fournier finished an easy first. All along the route crowds had cheered him, and hurled bouquets into the car, and wished him good speed; but in Berlin the assembled populace went nearly frantic at his appearance. Fournier was overwhelmed with flowers, laurel wreaths, and other offerings; dukes, duchesses, and the great people of the land pressed for presentations; he was the hero of the hour.

Thus ended what may be termed a peaceful invasion of Germany by the French. Among other things it had shown that over an immense stretch of country, over roads in places bad as only German roads can be, the automobile was able to maintain an average speed superior to that of the express trains running between Paris and Berlin;

also that, in spite of the large number of cars employed in the race, the accidents to the public were a negligible quantity. It should be mentioned that the actual time occupied by Fournier was 16 hours 5 minutes; that out of the 109 starters 47 reached Berlin; and that Osmont on a motor cycle finished only 3 hours and 10 minutes behind the winner.

In England such racing would be undesirable and impossible, owing to the crookedness of our roads. It would certainly not be permissible so long as the 12 miles an hour limit is observed. At the present time an agitation is on foot against this restriction, which, though reasonable enough among traffic and in towns, appears unjustifiable in open country. To help to convince the magisterial mind of the ease with which a car can be stopped, and therefore of its safety even at comparatively high speeds, trials were held on January 2, 1902, in Welbeck Park. The results showed that a car travelling at 13 miles an hour could be stopped dead in 4 yards; at 18 miles in 7 yards; at 20 miles in 13 yards; or in less than half the distance required to pull up a horse-vehicle driven at similar speeds.

Uses

Ninety-five per cent of motors, at least in England, are attached to pleasure vehicles, cycles, voiturettes, and large cars. On account of the costliness of cars motorists are far less numerous than cyclists; but those people whose means enable them to indulge in automobilism find it extremely fascinating. Caricaturists have presented to us in plenty the gloomier incidents of motoring; the broken chain, the burst tyre, the 'something gone wrong'. It requires personal experience to understand how lightly these mishaps weigh against the exhilaration of movement, the rapid change of scene, the sensation of control over power which can whirl one along tirelessly at a pace altogether beyond the capacities of horseflesh.

If proof were wanted of the motor car's popularity it will be seen in the unconventional dress of the chauffeur. The breeze set up by his rapid rush is such as would penetrate ordinary clothing; he dons cumbrous fur cloaks. The dust is all-pervading at times; he swathes himself in dust-proof overalls, and mounts large goggles edged with velvet, while a cap of semi-nautical cut tightly drawn down over neck and ears serves to protect those portions of his anatomy. The general effect is peculiarly unpicturesque; but even the most artistically-minded driver is ready to sacrifice appearances to comfort and the proper enjoyment of his car.

In England the great grievance of motorists arises from the speed limit imposed by law. To restrict a powerful car to twelve miles an hour is like confining a thoroughbred to the paces of a broken-down cab horse. Careless driving is unpardonable, but its occasional existence scarcely justifies the intolerant attitude of the law towards motorists in general. It must, however, be granted in justice to the police that the chauffeur, from constant transgression of the law, becomes a bad judge of speed, and often travels at a far greater velocity than he is willing to admit.

The convenience of the motor car for many purposes is immense, especially for cross-country journeys, which may be made from door to door without the monotony or indirectness of railway travel. It bears the doctor swiftly on his rounds. It carries the business man from his country house to his office. It delivers goods for the merchant; parcels for the post office.

In the warfare of the future, too, it will play its part, whether to drag heavy ordnance and stores, or to move commanding officers from point to point, or perform errands of mercy among the wounded. By the courtesy of the Locomobile Company we are permitted to append the testimony of Captain R. S. Walker, R.E., to the usefulness of a car during the great Boer War.

> 'Several months ago I noticed a locomobile car at Cape Town, and being struck with its simplicity and neatness, bought it and took it up country with me, with a view to making some tests with it over bad roads, etc. Its first trip was over a rough course round Pretoria, especially chosen to find out defects before taking it into regular use. Naturally, as the machine was not designed for this class of work, there were several. In about a month these had all been found out and remedied, and the car was in constant use, taking stores, etc., round the towns and

forts. It also performed some very useful work in visiting out-stations, where searchlights were either installed or wanted, and in this way visited nearly all the bigger towns in the Transvaal. It was possible to go round all the likely positions for a search-light in one day at every station, which frequently meant con-siderably over fifty miles of most indifferent roads, more than a single horse could have been expected to do, and the car gener-ally carried two persons on these occasions.

The car was also used as a tender to a searchlight plant, on a gun-carriage and limber, being utilised to fetch gasoline, car-bons, water, etc. etc., and also to run the dynamo for charging the accumulators used for sparking, thus saving running the gasoline motor for this purpose. To do this the trail of the car-riage, on which was the dynamo, was lowered on to the ground, the back of the car was pulled up, one wheel being supported on the dynamo pulley and the other clear of the ground, and two bolts were passed through the balance-gear to join it. On one occasion the car ran a 30 c.m. searchlight for an hour, driving a dynamo in this way. In consequence of this a trailer has been made to carry a dynamo and projector for search-lighting in the field, but so far this has not been so used. The trailer hooks into an eye, passing just behind the balance-gear. A Maxim, Colt, or small ammunition cart, etc., could be attached to this same eye.

Undoubtedly the best piece of work done by the car so far was its trial trip with the trailer, when it blew up the mines at Klein Nek. These mines were laid some eight months previously, and had never been looked to in the interval. There had been several bad storms, the Boers and cattle had been frequently through the Nek, it had been on fire, and finally it was shelled with lyd-dite. The mines, eighteen in number, were found to be intact except two, which presumably had been fired off by the heat of the veldt fire. All the insulation was burnt off the wires, and the battery was useless. It had been anticipated that a dynamo exploder would be inadequate to fire these mines, so a 250 volt two h.p. motor, which happened to be in Pretoria, weighing about three or four hundredweight, was placed on the trailer; a quarter of a mile of insulated cable, some testing gear, the kits of three men and their rations for three days, with a case of gasoline for the car, were also carried on the car and trailer, and the whole left Pretoria one morning and trekked to Rietfontein. Two of us were mounted, the third drove the car.

At Rietfontein we halted for the night, and started next morning with an escort through Commando Nek, round the north of the Magaliesburg, to near Klein Nek, where the road had to be left, and the car taken across country through bush veldt. At the bottom the going was pretty easy; only a few bushes had to be charged down, and the grass, etc., rather wound itself around the wheels and chain. As the rise became steeper the stones became very large, and the car had to be taken along very gingerly to prevent breaking the wheels.

A halt was made about a quarter of a mile from the top of the Nek, where the mines were. These were reconnoitred, and the wire, etc., was picked up; that portion which was useless was placed on top of the charges, and the remainder taken to the car. The dynamo was slid off the trailer, the car backed against it; one wheel was raised slightly and placed against the dynamo pulley, which was held up to it by a man using his rifle as a lever; the other wheel was on the ground with a stone under it. The balance gear being free, the dynamo was excited without the other wheel moving, and the load being on for a very short time (that is, from the time of touching lead on dynamo terminal to firing of the mine) no harm could come to the car. When all the leads had been joined to the dynamo the car was started, and after a short time, when it was judged to have excited, the second terminal was touched, a bang and clouds of dust resulted, and the Klein Nek Minefield had ceased to exist.

The day was extremely hot, and the work had not been light, so the tea, made with water drawn direct from the boiler, which we were able to serve round to the main body of our escort was much appreciated, and washed down the surplus rations we dispensed with to accommodate the battery and wire, which we could not leave behind for the enemy.

On the return journey we found this extra load too much for the car, and had great difficulty getting up to Commando Nek, frequently having to stop to get up steam, so these materials were left at the first blockhouse, and the journey home continued in comfort.

A second night at Rietfontein gave us a rest after our labour, and the third afternoon saw us on our way back to Pretoria. As luck would have it, a sandstorm overtook the car, which had a lively time of it. The storm began by blowing the sole occupant's hat off, so, the two mounted men being a long way behind, he shut off steam and chased his hat. In the meantime the wind increased, and the car sailed off on its own, and was only just caught in time to save a smash. Luckily the gale was in the right direction, for the fire was blown out, and it was impossible to light a match in the open. The car sailed into a poort on the outskirts of Pretoria, got a tow from a friendly cart through it, and then steamed home after the fire had been relit.

The load carried on this occasion (without the battery, etc.) must have been at least five hundredweight besides the driver, which, considering the car is designed to carry two on ordinary roads, and that these roads were by no means ordinary, was no mean feat. The car, as ordinarily equipped for trekking, carries the following: Blankets, waterproof sheets, etc., for two men; four planks for crossing ditches, bogs, stones, etc.; all necessary tools and spare parts, a day's supply of gasoline, a couple of telephones, and one mile of wire. In addition, on the trailer, if used for search-lighting: One 30 cm. projector, one automatic lamp for projector, one dynamo (100 volts 20 amperes), two short lengths of wire, two pairs of carbons, tools, etc. This trailer would normally be carried with the baggage, and only picked up by the car when wanted as a light; that is, as a rule, after arriving in camp, when a good many other things could be left behind.'

Perhaps the most useful work in store for the motor is to help relieve the congestion of our large towns and to restore to the country some of its lost prosperity. There is no stronger inducement to

make people live in the country than rapid and safe means of locomotion, whether public or private. At present the slow and con-
gested suburban train services on some sides of London consume
as much time as would suffice a motor car to cover twice or three
times the distance. We must welcome any form of travel which will
tend to restore the balance between country and town by enabling
the worker to live far from his work. The gain to the health of the
nation arising from more even distribution of population would be
inestimable.

A world's tour is among the latest projects in automobilism. On
April 29, 1902, Dr. Lehwess and nine friends started from Hyde
Park Corner for a nine months' tour on three vehicles, the larg-
est of them a luxuriously appointed 24 horsepower caravan, built
to accommodate four persons. Their route lies through France,
Germany, Russia, Siberia, China, Japan, and the United States.

Chapter 8

High-speed Railways

A century ago a long journey was considered an exploit, and an exploit to be carried through as quickly as possible on account of the dangers of the road and the generally uncomfortable conditions of travel. To-day, though our express speed is many times greater than that of the lumbering coaches, our carriages comparatively luxurious, the risk practically nil, the same wish lurks in the breast of ninety-nine out of a hundred railway passengers; to spend the shortest time in the train that the time-table permits. Time differences that to our grandfathers would have appeared trifling are now matters of sufficient importance to make rival railway companies anxious to clip a few minutes off a 100 mile 'run' simply because their passengers appreciate a few minutes' less confinement to the cars.

During the last fifty years the highest express speeds have not materially altered. The Great Western Company in its early days ran trains from Paddington to Slough, 18 miles, in 15.5 minutes, or at an average pace of 69.5 miles an hour.

On turning to the present regular express services of the world we find America heading the list with a 50-mile run between Atlantic City and Camden, covered at the average speed of 68 miles an hour; Britain second with a 33 mile run between Forfar and Perth at 59 miles an hour; and France a good third with an hourly average of rather more than 58 miles an hour between Lee Arbutus and S. Pierre des Corps. These runs are longer than that on the Great Western Railway referred to above (which now occupies twenty-four minutes), but their average velocity is less. What is the cause of this decrease of speed? Not want of power in modern engines; at times our trains attain a rate of 80 miles an hour, and in America a mile has been turned off in the astonishing time of thirty-two seconds. We should rather seek it in the need for economy and in the physical limitations imposed by the present system of plate laying and railroad engineering. An average speed of ninety miles an hour would, as things now stand, be too wasteful of coal and too injurious to the rolling-stock to yield profit to the proprietors of a line; and, except in certain districts, would prove perilous for the passengers. Before our services can be much improved the steam locomotive must be supplanted by some other application of motive power, and the metals be laid in a manner which will make special provision for extreme speed.

Since rapid transit is as much a matter of commercial importance as of mere personal convenience it must not be supposed that an average of 50 miles an hour will continue to meet the needs of travellers. Already practical experiments have been made with two systems that promise us an ordinary speed of 100 miles an hour and an express speed considerably higher.

The Monorail

One of these, the monorail or single-rail system, will be employed on a railroad projected between Manchester and Liverpool. At present passengers between these two cities, the first to be connected by a railroad of any kind, enjoy the choice of three rival services covering 34.5 miles in three quarters of an hour. An eminent engineer, Mr. F. B. Behr, now wishes to add a fourth of unprecedented swiftness. Parliamentary powers have been secured for a line starting from Deansgate, Manchester, and terminating behind the pro-Cathedral in Liverpool, on which single cars will run every ten minutes at a velocity of 110 miles an hour.

A monorail track presents a rather curious appearance. The ordinary parallel metals are replaced by a single rail carried on the summit of A-shaped trestles, the legs of which are firmly bolted to sleepers. A monorail car is divided lengthwise by a gap that allows it to hang half on either side of the trestles and clear them as it moves. The double flanged wheels to carry and drive the car are placed at the apex of the gap. As the centre of gravity is below the rail, the car cannot turn over, even when travelling round a sharp curve.

The first railway built on this system was constructed by M. Charles Latrine, a French engineer, in Algeria, a district where an ordinary two-rail track is often blocked by severe sand-storms. He derived the idea of balancing trucks over an elevated rail from caravans of camels laden on each flank with large bags. The camel, or rather its legs, was transformed by the engineer's eye into iron trestles, while its burden became a car. A line built as a result of this observation, and supplied with mules as tractive power, has for many years played an important part in the esparto-grass trade of Algeria.

In 1886 Mr. Behr decided that by applying steam to M. Latrine's system he could make it successful as a means of transporting passengers and goods. He accordingly set up in Toothily Fields, Westminster, on the site of the new Roman Catholic Cathedral, a miniature railway which during nine months of use showed that the monorail would be practical for heavy traffic, safe, and more cheaply maintained than the ordinary double-metal railway. The train travelled easily round very sharp curves and climbed unusually steep gradients without slipping.

Mr. Behr was encouraged to construct a monorail in Kerry, between Listowel, a country town famous for its butter, and Ballybunion, a seaside resort of increasing popularity. The line, opened on the 28th of February 1888, has worked most satisfactorily ever since, without injury to a single employee or passenger.

On each side of the trestles, two feet below the apex, run two guide-rails, against which press small wheels attached to the carriages to prevent undue oscillation and tipping round curves. At the three stations there are, instead of points, turn-tables or switches on to which the train runs for transference to sidings.

Road traffic crosses the rail on drawbridges, which are very easily worked, and which automatically set signals against the train. The bridges are in two portions and act on the principle of the Tower Bridge, each half falling from a perpendicular position towards the centre, where the ends rest on the rail, specially strengthened at that spot to carry the extra weight. The locomotive is a twill affair; has two boilers, two funnels, two fireboxes; can draw 240 tons on the level at fifteen miles all hour, and when running light travels a mile in two minutes. The carriages, 18 feet long and carrying twelve passengers on each side, are divided longitudinally into two parts. Trucks too are used, mainly for the transport of sand, of which each carries three tons, from Ballybunion to Listowel. In the centre of each train is a queer-looking vehicle serving as a bridge for any one who may wish to cross from one side of the rail to the other.

Several lines on the pattern of the Ballybunion Listowel have been erected in different countries. Mr. Behr was not satisfied with his first success, however, and determined to develop the monorail in the direction of fast travelling, which he thought would be most easily attained on a trestle-track. In 1893 he startled engineers by proposing a Lightning Express service, to transport passengers at a velocity of 120 miles an hour. But the project seemed too ideal to tempt money from the pockets of financiers, and Mr. Behr soon saw that if a high-speed railway after his own heart were constructed it must be at his own expense. He had sufficient faith in his scheme to spend £40,000 on an experimental track at the Brussels Exhibition of 1897. The exhibition was in two parts, connected by an electric railway, the one at the capital, the other at Tervueren, seven miles away. Mr. Behr built his line at Tervueren.

The greatest difficulty he encountered in its construction arose from the opposition of landowners, mostly small peasant proprietors, who were anxious to make advantageous terms before they would hear of the rail passing through their lands. Until he had concluded two hundred separate contracts, by most of which the peasants benefited, his plate layers could not get to work. Apart from this opposition the conditions were not favourable. He was obliged to bridge no less than ten roads; and the contour of the country necessitated steep gradients, sharp curves, long cuttings and embankments, the last of which, owing to a wet summer, could not be trusted to stand quite firm. The track was doubled for three miles, passing at each end round a curve of 1,600 feet radius.

The rail ran about four feet above the track on trestles bolted down to steel sleepers resting on ordinary ballast. The carriage Mr. Behr used but one on this line-weighed 68 tons, was 59 feet long and 11 feet wide, and could accommodate one hundred persons. It was handsomely fitted up, and had specially-shaped seats which neutralised the effect of rounding curves, and ended fore and aft in a point, to overcome the wind-resistance in front and the air suction behind. Sixteen pairs of wheels on the under side of the carriage engaged with the two pairs of guide rails flanking the trestles, and eight large double-flanged wheels, 4.5 feet in diameter, carried the weight of the vehicle. The inner four of these wheels were driven by as many powerful electric motors contained, along with the guiding mechanism, in the lower part of the car. The motors picked up current from the centre rail and from another steel rail laid along the sleepers on porcelain insulators.

The top speed attained was about ninety miles an hour. On the close of the Exhibition special experiments were made at the request of the Belgian, French, and Russian Governments, with results that proved that the Behr system deserved a trial on a much larger scale.

The engineer accordingly approached the British Government with a Bill for the construction of a high-speed line between Liverpool and Manchester. A Committee of the House of Commons rejected the Bill on representations of the Salford Corporation. The Committee had to admit, nevertheless, that the evidence called was mainly in favour of the system; and, the plans of the rail having been altered to meet certain objections, Parliamentary consent was obtained to commence operations when the necessary capital had been subscribed. In a few years the great seaport and the great cotton town will probably be within a few minutes' run of each other.

A question that naturally arises in the mind of the reader is this; could the cars, when travelling at 110 miles an hour, be arrested quickly enough to avoid an accident if anything got on the line?

The Westinghouse air-brake has a retarding force of three miles a second. It would therefore arrest a train travelling at 110 miles per hour in 37 seconds, or 995 yards. Mr. Behr proposes to reinforce the Westinghouse with an electric brake, composed of magnets 18 inches long, exerting on the guide rails by means of current generated by the reversed motors an attractive force of 200 Ibs, per square inch. One great advantage of this brake is that its efficiency is greatest when the speed of the train is highest and when it is most needed. The united brakes are expected to stop the car in half the distance of the Westinghouse alone; but they would not both be applied except in emergencies. Under ordinary conditions the slowing of a car would take place only at the termini, where the line ascends gradients into the stations. There would, however, be small chance of collisions, the railway being securely fenced off throughout its entire length, and free from level crossings, drawbridges and points. Furthermore, each train would be its own signal-man. Suppose the total 34.5 miles divided into 'block' lengths of 7 miles. On leaving a terminus the train sets a danger signal behind it; at 7 miles it sets another, and at 14 miles releases the first signal. So that the driver of a car would have at least 7 miles to slow down in after seeing the signals against him. In case of fog he would consult a miniature signal in his cabin working electrically in unison with the large semaphores.

The Manchester-Liverpool rail will be reserved for express traffic only. Mr. Behr does not believe in mixing speeds, and considers it one of the advantages of his system that slow cars and wagons of the ordinary two-rail type cannot be run on the monorail; because if they could managers might be tempted to place them there.

A train will consist of a single vehicle for forty, fifty, or seventy passengers, as the occasion requires. It is calculated that an average of twelve passengers at one penny per mile would pay all the expenses of running a car.

Mr. Behr maintains that monorails can be constructed far more cheaply than the two-rail, because they permit sharper curves, and thereby save a lot of cutting and embankment; and also because the monorail itself, when trestles and rail are specially strengthened, can serve as its own bridge across roads, valleys and rivers.

Electric Locomotive Railways

Though the single-rail has come to the front of late, it must not be supposed that the two-rail track is forever doomed to moderate speeds only. German engineers have built an electric two-rail military line between Berlin and Zossen, seventeen miles long, over which cars have been run at a hundred miles an hour. The line has very gradual curves, and in this respect is inferior to the more sinuous monorail. Its chief virtue is the method of applying motive power, a method common to both systems.

The steam locomotive creates its own motive force, and as long as it has fuel and water can act independently. The electric locomotive, on the other hand, receives its power through metallic conductors from some central station. Should the current fail all the traffic on the line is suspended. So far the advantage rests with the steamer. But as regards economy the superiority of the current is obvious.

In the electric systems under consideration, the monorail and Berlin-Zossen, there is less weight per passenger to be shifted, since a comparatively light motor supersedes the heavy locomotive. The cars running singly, bridges and track are subjected to less strain, and cost less to keep in repair. But the greatest saving of all is made in fuel. A steam locomotive uses coal wastefully, sending a lot of latent power up the funnel in the shape of half-expanded steam. Want of space prevents the designer from fitting to a moving engine the more economical machinery to be found in the central power-station of an electric railway, which may be so situated-by the water-side or near a pit's mouth-that fuel can be brought to it at a trifling cost. Not only is the expense of distributing coal over the system avoided, but the coal itself, by the help of triple and quadruple expansion engines should yield two or three times as much energy per ton as is developed in a locomotive furnace.

Many schemes are afoot for the construction of high-speed railways. The South-Eastern plans a monorail between Cannon Street and Charing Cross to avoid the delay that at present occurs in passing from one station to the other. We hear also of a projected railway from London to Brighton, which will reduce the journey to half-an-hour; and of another to connect Dover and London. It has even been suggested to establish monorails on existing tracks for fast passenger traffic, the expresses passing overhead, the slow and goods trains plodding along the double metals below.

But the most ambitious programme of all comes from the land of the Czar. M. Hippolyte Romanoff, a Russian engineer, proposes to unite St. Petersburg and Moscow by a line that shall cover the intervening 600 miles in three hours, an improvement of ten hours on the present time-tables. He will use T-shaped supports to carry two rails, one on each arm, from which the cars are to hang. The line being thus do-able will permit the cars, some four hundred in number, to run to and fro continuously, urged on their way by current picked up from overhead wires. Each car is to have twelve wheels, four drivers arranged vertically and eight horizontally, to prevent derailment by gripping the rail on either side. The stoppage or breakdown of any car will automatically stop those following by cutting off the current.

In the early days of railway history lines were projected in all directions, regardless of the fact whether they would be of any use or not. Many of these lines began, where they ended, on paper. And now that the high-speed question has cropped up, we must not believe that every projected electric railway will be built, though of the ultimate prevalence of far higher speeds than we now enjoy there can be doubt.

What would become of the records established in the 'Race to the North' and by American 'fliers'. And what about continental travel?

Assuming that the Channel Tunnel is built – perhaps a rather large assumption – Paris will be at our very doors. A commercial traveller will step into the lightning express at London, sleep for two hours and twenty-four minutes and wake, refreshed, to find the blue-smocked Paris porters bawling in his ear. Or even if we prefer to keep the 'little silver streak' free from subterranean burrows, he will be able to catch the swift turbine steamers at Dover, slip across to Calais in half-an-hour, and be at the French capital within four hours of quitting London. And if M. Romanies' standard be reached, the latest thing in hats despatched from Paris at noon may be worn in Regent Street before two o'clock.

Such speeds would indeed produce a revolution in travelling comparable to the substitution of the steam locomotive for the stage coach. As has been pithily said, the effect of steam was to make the bulk of population travel, whereas they had never travelled before, but the effect of the electric railway will be to make those who travel go further and more frequently.

Chapter 9

The Sea Express

In the year 1836 the Sirius, a paddle-wheel vessel, crossed the Atlantic from Cork Harbour to New York in nineteen days. Contrast with the first steam-passage from the Old World to the New a return journey of the Deutschland, a North German liner, which in 1900 averaged over twenty-seven miles an hour between Sandy Hook and Plymouth, accomplishing the whole distance in the record time of five days seven hours thirty-eight minutes.

This growth of speed is even more remarkable than might appear from the mere comparison of figures. A body moving through water is so retarded by the inertia and friction of the fluid that to quicken its pace a force quite out of proportion to the increase of velocity must be exerted. The proportion cannot be reduced to an exact formula, but under certain conditions the speed and the power required advance in the ratio of their cubes; that is, to double a given rate of progress eight times the driving-power is needed; to treble it, twenty-seven times.

The mechanism of our fast modern vessels is in every way as superior to that which moved the Sirius, as the beautifully-adjusted safety cycle is to the clumsy 'bone-shaker' which passed for a wonder among our grandfathers. A great improvement has also taken place in the art of building ships on lines calculated to offer least resistance to the water, and at the same time afford a good carrying capacity. The big liner, with its knife-edged bow and tapering hull, is, by its shape alone, eloquent of the high speed which has earned it the title of Ocean Greyhound; and as for the fastest craft of all, torpedo-destroyers, their designers seem to have kept in mind Euclid's definition of a line; length without breadth. But whatever its shape, boat or ship may not shake itself free of Nature's laws. Her restraining hand lies heavy upon it. A single man paddles his weight carrying dinghy along easily at four miles an hour; eight men in the pink of condition, after arduous training, cannot urge their light, slender, racing shell more than twelve miles in the same time.

Inside the Sea Express

To understand how mail boats and destroyers attain, despite the enormous resistance of water, velocities that would shame many a train-service, we have only to visit the stoke-holds and engine-rooms of our sea expresses and note the many devices of marine engineers by which fuel is converted into speed.

We enter the stoke-hold through air-locks, closing one door before we can open the other, and find ourselves among sweating, grimy men, stripped to the waist. As though life itself depended upon it they shovel coal into the rapacious maws of furnaces glowing with a dazzling glare under the 'forced draught' sent down into the hold by the fans whirling overhead. The ignited furnace gases on their way to the outer air surrender a portion of their heat to the water from which they are separated by a skin of steel. Two kinds of marine boiler are used; the fire-tube and the water-tube. In fire-tube boilers the fire passes inside the tubes and the water outside; in water-tube boilers the reverse is the case, the crown and sides of the furnace being composed of sheaves of small parallel pipes through which water circulates. The latter type, as generating steam very quickly, and being able to bear very high pressures, is most often found in war vessels of all kinds. The quality sought in boiler construction is that the heating surface should be very large in proportion to the quantity of water to be heated. Special coal, anthracite or Welsh, is used in the navy on account of its great heating power and freedom from smoke; experiments have also been made with crude petroleum, or liquid fuel, which can be more quickly put on board than coal, requires the services of fewer stokers, and may be stored in odd corners unavailable as coal bunkers.

From the boiler the steam passes to the engine-room, whither we will follow it. We are now in a bewildering maze of clanking, whirling machinery; our noses offended by the reek of oil, our ears deafened by the uproar of the moving metal, our eyes wearied by the efforts to follow the motions of the cranks and rods.

On either side of us is ranged a series of three or perhaps even four cylinders, of increasing size. The smallest, known as the high-pressure cylinder, receives steam direct from the boiler. It takes in through a slide-valve a supply for a stroke; its piston is driven from end to end; the piston-rod flies through the cylinder-end and transmits a rotary motion to a crank by means of a connecting-rod. The half-expanded steam is then ejected, not into the air as would happen on a locomotive, but into the next cylinder, which has a larger piston to compensate the reduction of pressure. Number two served, the steam does duty a third time in number three, and perhaps yet a fourth time before it reaches the condensers, where its sudden conversion into water by cold produces a vacuum suction in the last cylinder of the series. The secret of a marine engine's strength and economy lies then in its treatment of the steam, which, like clothes in a numerous family, is not thought to have served its purpose till it has been used over and over again.

Reciprocating (i.e. cylinder) engines, though brought to a high pitch of efficiency, have grave disadvantages, the greatest among which is the annoyance caused by their intense vibration to all persons in the vessel. A revolving body that is not exactly balanced runs unequally, and transmits a tremor to anything with which it may be in contact. Turn a cycle upside down and revolve the driving-wheel rapidly by means of the pedal. The whole machine soon begins to tremble violently, and dance up and down on the saddle springs, because one part of the wheel is heavier than the rest, the mere weight of the air-valve being sufficient to disturb the balance.

Now consider what happens in the engine-room of high-powered vessels. On destroyers the screws make 400 revolutions a minute. That is to say, all the momentum of the pistons, cranks, rods, and valves (weighing tons), has to be arrested thirteen or fourteen times every second. However well the moving parts may be balanced, the vibration is felt from stem to stern of the vessel. Even on luxuriously-appointed liners, with engines running at a far slower speed, the throbbing of the screw (i.e. engines) is only too noticeable and productive of discomfort.

We shall be told, perhaps, that vibration is a necessary consequence of speed. This is true enough of all vehicles, such as railway trains, motor-cars, cycles, which are shaken by the irregularities of the unyielding surface over which they run, but does not apply universally to ships and boats. A sail or oar-propelled craft may be entirely free from vibration, whatever its speed, as the motions arising from water are usually slow and deliberate. In fact, water in its calmer moods is an ideal medium to travel on, and the trouble begins only with the introduction of steam as motive force.

The Steam Turbine

But even steam may be robbed of its power to annoy us. The steam-turbine has arrived. It works a screw propeller as smoothly as a dynamo, and at a speed that no cylinder engine could maintain for a minute without shaking itself to pieces.

The steam-turbine is most closely connected with the name of the Hon. Charles Parsons, son of Lord Rosse, the famous astronomer. He was the first to show, in his speedy little Turbinia, the possibilities of the turbine when applied to steam navigation. The results have been such as to attract the attention of the whole shipbuilding world.

The principle of the turbine is seen in the ordinary windmill. To an axle revolving in a stationary bearing are attached vanes which oppose a current of air, water, or steam, at an angle to its course, and by it are moved sideways through a circular path. Mr. Parsons' turbine has of course been specially adapted for the action of steam.

It consists of a cylindrical, air-tight chest, inside which rotates a drum, fitted round its circumference with rows of curved vanes. The chest itself has fixed immovably to its inner side a corresponding number of vane rings, alternating with those on the drum, and so arranged as to deflect the steam on to the latter at the most efficient angle. The diameter of the chest and drum is not constant, but increases towards the exhaust end, in order to give the expanding and weakening steam a larger leverage as it proceeds.

The steam entering the chest from the boiler at a pressure of some hundreds of pounds to the square inch strikes the first set of vanes on the drum, passes them and meets the first set of chest-vanes, is turned from its course on to the second set of drum-vanes, and so on to the other end of the chest. Its power arises entirely from its expansive velocity, which, rather than turn a number of sharp corners, will, if possible, compel the obstruction to move out of its way. If that obstruction be from any cause difficult to stir, the steam must pass round it until its pressure overcomes the inertia. Consequently the turbine differs from the cylinder engine in this respect, that steam can pass through and be wasted without doing any work at all, whereas, unless the gear of a cylinder moves, and power is exerted, all steam ways are closed, and there is no waste. In practice, therefore, it is found that a turbine is most effective when running at high speed.

The first steam-turbines were used to drive dynamos. In 1884 Mr. Parsons made a turbine in which fifteen wheels of increasing size moved at the astonishing rate of 300 revolutions per second, and developed 10 horse-power. In 1888 followed a 120 horse-power turbine, and in 1892 one of 2,000 horse-power, provided with a condenser to produce suction. So successful were these steam fans for electrical work, pumping water and ventilating mines, that Mr. Parsons determined to test them as a means of propelling ships. A small vessel 100 feet long and 9 feet in beam was fitted with three turbines, high, medium, and low pressure, of a total 2,000 horse-power, a proportion of motive force to tonnage hitherto not approached. Yet when tried over the test course the Turbinia, as the boat was fitly named, ran in a most disappointing fashion. The screws revolved too fast, producing what is known as cavitation, or the scooping out of the water by the screws, so that they moved in a partial vacuum and utilised only a fraction of their force, from lack of anything to 'bite' on. This defect was remedied by employing screws of coarser pitch and larger blade area, three of which were attached to each of the three propeller shafts. On a second trial the Turbinia attained 32.75 knots over the 'measured mile', and later the astonishing speed of forty miles an hour, or double that of the fast Channel packets. At the Spithead Review in 1897 one of the most interesting sights was the little nimble Turbinia rushing up and down the rows of majestic warships at the rate of an express train.

After this success Mr. Parsons erected works at Wallsend-on-Tyne for the special manufacture of turbines. The Admiralty soon placed with him an order for a torpedo-destroyer; the Viper, of 350 tons; which on its trial trip exceeded forty-one miles an hour at an estimated horse-power (11,000) equalling that of our largest battleships. A sister vessel, the Cobra, of like size, proved as speedy. Misfortune, however, overtook both destroyers. The Viper was wrecked August 3, 1901, on the coast of Alderney during the Autumn naval manoeuvres, and the Cobra foundered in a severe storm in September of the same year in the North Sea. This double disaster casts no reflections on the turbine engines; being attributed to fog in the one case and to structural weakness in the other. The Admiralty has since ordered another turbine destroyer, and before many years are past we shall probably see all the great naval powers providing themselves with like craft to act as the 'eyes of the fleet', and travel at even higher speeds than those of the Viper and Cobra.

The turbine has been applied to mercantile as well as warlike purposes. There is at the present time a turbine-propelled steamer, the King Edward, running in the Clyde on the Fairlie-Campbelltown route. This vessel, 250 feet long, 30 broad, 18 deep, contains three turbines. In each the steam is expanded fivefold, so that by the time it passes into the condensers it occupies 125 times its boiler volume. (On the Viper the steam entered the turbine through an inlet eight inches in diameter, and left them by an outlet four feet square.) In cylinder engines thirty-fold expansion is considered a high ratio;

hence the turbine extracts a great deal more power in proportion from its steam. As a turbine cannot be reversed, special turbines are attached to the two outside of the three propeller shafts to drive the vessel astern. The steamer attained 20 ½ knots over the 'Skelmorlie mile' in fair and calm weather, with 3,500 horse-power produced at the turbines. The King Edward is thus the fastest by two or three knots of all the Clyde steamers, as she is the most comfortable. We are assured that as far as the turbines are concerned it is impossible by placing the hand upon the steam-chest to tell whether the drum inside is revolving or not!

Every marine engine is judged by its economy in the consumption of coal. Except in times of national peril extra speed produced by an extravagant use of fuel would be severely avoided by all owners and captains of ships. At low speeds the turbine develops less power than cylinders from the same amount of steam, but when working at high velocity it gives at least equal results. A careful record kept by the managers of the Caledonian Steamship Company compares the King Edward with the Duchess of Hamilton, a paddle steamer of equal tonnage used on the same route and built by the same firm. The record shows that though the paddle-boat ran a fraction of a mile further for every ton of coal burnt in the furnaces, the King Edward averaged two knots an hour faster, a superiority of speed quite out of proportion to the slight excess of fuel. Were the Duchess driven at 18.5 knots instead of 16.5 her coal bill would far exceed that of the turbine.

As an outcome of these first trials the Caledonian Company are launching a second turbine vessel. Three high-speed turbine yachts are also on the stocks; one of 700 tons, another of 1,500 tons, and a third of 170 tons. The last, the property of Colonel M'Calmont, is designed for a speed of twenty-four knots.

Mr. Parsons claims for his system the following advantages: Greatly increased speed; increased carrying power of coal; economy in coal consumption; increased facilities for navigating shallow waters; greater stability of vessels; reduced weight of machinery (the turbines of the King Edward weigh but one-half of cylinders required to give the same power); cheapness of attending the machinery;

absence of vibration, lessening wear and tear of the ship's hull and assisting the accurate training of guns; lowered centre of gravity in the vessel, and consequent greater safety during times of war.

The inventor has suggested a cruiser of 2,800 tons, engined up to 80,000 horse-power, to yield a speed of forty-four knots (about fifty miles) an hour. Figures such as these suggest that we may be on the eve of a revolution of ocean travel comparable to that made by the substitution of steam for wind power. Whether the steam-turbine will make for increased speed all round, or for greater economy, remains to be seen; but we may be assured of a higher degree of comfort. We can easily believe that improvements will follow in this as in other mechanical contrivances, and that the turbine's efficiency has not yet reached a maximum; and even if our ocean expresses, naval and mercantile, do not attain the one-mile-a-minute standard, which is still regarded as creditable to the fastest methods of land locomotion, we look forward to a time in the near future when much higher speeds will prevail, and the tedium of long voyages be greatly shortened. Already there is talk of a service which shall reduce the trans-Atlantic journey to three-and-a-half days. The means are at hand to make it a fact.

Chapter 10
Mechanical Flight

Few, if any, problems have so strongly influenced the imagination and exercised the ingenuity of mankind as that of aerial navigation. There is something in our nature that rebels against being condemned to the condition of 'featherless bipeds' when birds, bats, and even minute insects have the whole realm of air and the wide heavens open to them. Who has not, like Solomon, pondered upon 'the way of a bird in the air' with feelings of envy and regret that he is chained to earth by his gross body; contrasting our laboured movements from point to point of the earth's surface with the easy gliding of the feathered traveller? The unrealised wish has found expression in legends of Daedalus, Pegasus, in the 'flying carpet' of the fairy tale, and in the pages of Jules Verne, in which last the adventurous Robur on his 'Clipper of the Clouds' anticipates the future in a most startling fashion.

Aeromobilism, to use its most modern title is regarded by the crowd as the mechanical counterpart of the Philosopher's Stone or the Elixir of Life; a highly desirable but unattainable thing. At times this incredulity is transformed by highly-coloured press reports into an equally unreasonable readiness to believe that the conquest of the air is completed, followed by a feeling of irritation that facts are not as they were represented in print.

The proper attitude is of course half-way between these extremes. Reflection will show us that money, time, and life itself would not have been freely and ungrudging given or risked by many men, hard-headed, practical men among them, in pursuit of a Will-o'-the-Wisp, especially in a century when scientific calculation tends always to calm down any too imaginative scheme. The existing state of the aerial problem may be compared to that of a railway truck which an insufficient number of men are trying to move. Ten men may make no impression on it, though they are putting out all their strength. Yet the arrival of an eleventh may enable them to overcome the truck's inertia and move it at an increasing pace.

Every new discovery of the scientific application of power brings us nearer to the day when the truck will move. We have metals of wonderful strength in proportion to their weight; pigmy motors containing the force of giants; a huge fund of mechanical experience to draw upon; in fact, to paraphrase the jingo song, 'We've got the things, we've got the men, we've got the money too', but we haven't as yet got the machine that can mock the bird like the flying express mocks the strength and speed of horses.

The reason of this is not far to seek. The difficulties attending the creation of a successful flying-machine are immense, some unique, not being found in aquatic and terrestrial locomotion.

In the first place, the airship, flying-machine, aerostat, or whatever we please to call it, must not merely move, but also lift itself. Neither a ship nor a locomotive is called upon to do this. Its ability to lift itself must depend upon either the employment of large balloons or upon sheer power. In the first case the balloon will, by reason of its size, be unmanageable in a high wind; in the second case, a breakdown in the machinery would probably prove fatal.

Even supposing that our aerostat can lift itself successfully, we encounter the difficulties connected with steering in a medium traversed by ever-shifting currents of air, which demands of the helmsman a caution and capacity seldom required on land or water. Add to these the difficulties of leaving the ground and alighting safely upon it; and, what is more serious than all, the fact that though success can be attained only by experiment, experiment is in this case extremely expensive and risky, any failure often resulting in total ruin of the machine, and sometimes in loss of life. The list of those who have perished in the search for the power of flight is a very long one.

Yet in spite of these obstacles determined attempts have been and are being made to conquer the air. Men in a position to judge are confident that the day of conquest is not very far distant, and that the next generation may be as familiar with aerostats as we with motor-cars. Speculation as to the future is, however, here less profitable than a consideration of what has been already done in the direction of collecting forces for the final victory.

To begin at the beginning, we see that experimenters must be divided into two great classes: those who pin their faith to airships lighter than air, e.g. Santos Dumont, Zeppelin, Roze; and those who have small respect for balloons, and see the ideal air-craft in a machine lifted entirely by means of power and surfaces pressing the air after the manner of a kite. Sir Hiram Maxim and Professor S. P. Langley, Mr. Lawrence Hargrave, and Mr. Sydney Hollands are eminent members of the latter cult.

Steerable Balloons

As soon as we get on the topic of steerable balloons the name of Mr. Santos Dumont looms large. But before dealing with his exploits we may notice the airship of Count Zeppelin, an ingenious and costly structure that was tested over Lake Constance in 1900.

The balloon was built in a large wooden shed 450 by 78 by 66 feet, that floated on the lake on ninety pontoons. The shed alone cost over 10,000 pounds.

The balloon itself was nearly 400 feet long, with a cylindrical diameter of 39 feet, except at its ends, which were conical, to offer as little resistance as possible to the air. Externally it afforded the appearance of a single-compartment bag, but in reality it was divided into seventeen parts, each gas-tight, so that an accident to one part of the fabric should not imperil the whole. A framework of aluminium rods and rings gave the bag a partial rigidity.

Its capacity was 12,000 cubic yards of hydrogen gas, which, as our readers doubtless know, is much lighter though more expensive than ordinary coal-gas; each inflation costing several hundreds of pounds. Under the balloon hung two cars of aluminium, the motors and the screws; and also a great sliding weight of 600 pounds for altering the 'tip' of the airship; and rudders to steer its course.

On June 30th a great number of scientific men and experts assembled to witness the behaviour of a balloon which had cost 20,000 pounds. For two days wind prevented a start, but on July 2nd, at 7.30 p.m., the balloon emerged from its shed, and at eight o'clock commenced its first journey, with and against a light easterly wind for a distance of three and a half miles. A mishap to the steering-gear occurred early in the trip, and prevented the airship appearing to advantage, but a landing was effected easily and safely. In the following October the Count made a second attempt, returning against a wind blowing at three yards a second, or rather more than six miles an hour. Owing to lack of funds the fate of the 'Great Eastern' has overtaken the Zeppelin airship; to be broken up, and the parts sold.

The aged Count had demonstrated that a petroleum motor could be used in the neighbourhood of gas without danger. It was, however, reserved for a younger man to give a more decided proof of the steerableness of a balloon.

In 1900 M. Henry Deutsch, a member of the French Aerie Club, founded a prize of 4,000 pounds, to win which a competitor must start from the Aerie Club Park, near the Seine in Paris, sail to and round the Eiffel Tower, and be back at the starting-point within a time-limit of half-an-hour.

M. Santos Dumont, a wealthy and plucky young Brazilian, had, previously to this offer, made several successful journeys in motor balloons in the neighbourhood of the Eiffel Tower. He therefore determined to make a bid for the prize with a specially constructed balloon 'Santos Dumont V'. The third unsuccessful attempt ended in disaster to the airship, which fell on to the houses, but fortunately without injuring its occupant.

Another balloon; 'Santos Dumont VI' was then built. On Saturday, October 19th, M. Dumont reached the Tower in nine minutes and recrossed the starting line in 20.5 more minutes, thus complying with the conditions of the prize with half-a minute to spare. A dispute, however, arose as to whether the prize had been actually won, some of the committee contending that the balloon should have come to earth within the half-hour, instead of merely passing overhead; but finally the well-merited prize was awarded to the determined young aeronaut.

The successful airship was of moderate proportions as compared with that of Count Zeppelin. The cigar-shaped bag was 112 feet long and 20 feet in diameter, holding 715 cubic yards of gas. M. Dumont showed originality in furnishing it with a smaller balloon inside, which could be pumped full of air so as to counteract any leakage in the external bag and keep it taut. The motor, on which everything depended, was a four-cylinder, petrol-driven engine, furnished with 'water-jackets' to prevent over-heating. The motor turned a large screw – made of silk and stretched over light frames – 200 times a minute, giving a driving force of 175 pounds. Behind, a rudder directed the airship, and in front hung down a long rope suspended by one end that could be drawn towards the centre of the frame to alter the trim of the ship. The aeronaut stood in a large wicker basket flanked on either side by bags of sand ballast. The fact that the motor, once stopped, could only be restarted by coming to earth again added an element of great uncertainty to all his trips; and on one occasion the mis-firing of one of the cylinders almost brought about a collision with the Eiffel Tower.

From Paris M. Dumont went to Monaco at the invitation of the prince of that principality, and cruised about over the bay in his balloon. His fresh scheme was to cross to Corsica, but it was brought to an abrupt conclusion by a leakage of gas, which precipitated balloon and balloonist into the sea. Dumont was rescued, and at once set about new projects, including a visit to the Crystal Palace, where he would have made a series of ascents this summer (1902) but for damage done to the silk of the gas-bag by its immersion in salt water and the other vicissitudes it had passed through. Dumont's most important achievement has been, like that

of Count Zeppelin, the application of the gasoline motor to aeromobilism. In proportion to its size this form of motor develops a
large amount of energy, and its mechanism is comparatively simple, a matter of great moment to the aeronaut. He has also shown
that under favourable conditions a balloon may be steered against a
head-wind, though not with the certainty that is desirable before air
travel can be pronounced an even moderately simple undertaking.

The Aeroplane

The fact that many inventors, such as Dr. Barton, M. Roze, Henri
Deutsch, are fitting motors to balloons in the hopes of solving the
aerial problem shows that the airship has still a strong hold on the
minds of men. But on reviewing the successes of such combinations of lifting and driving power it must be confessed, with all due
respect to M. Dumont, that they are somewhat meagre, and do not
show any great advance.

The question is whether these men are not working on wrong lines,
and whether their utmost endeavours and those of their successors
will ever produce anything more than a very semi-successful craft.
Their efforts appear foredoomed to failure. As Sir Hiram Maxim
has observed, a balloon by its very nature is light and fragile, it is a
mere bubble. If it were possible to construct a motor to develop 100
horsepower for every pound of its weight, it would still be impossible to navigate a balloon against a wind of more than a certain
strength. The mere energy of the motor would crush the gas-bag
against the pressure of the wind, deform it, and render it unmanageable. Balloons therefore must be at the mercy of the wind, and
obliged to submit to it under conditions not always in accordance
with the wish of the aeronaut.

Sir Hiram in condemning the airship was ready with a substitute.
On looking round on the patterns of nature he concluded that, inasmuch as all things that fly are heavier than air, the problem of aerial
navigation must be solved by a machine whose natural tendency is
to fall to the ground, and which can be sustained only by the exer-
tion of great force. Its very weight would enable it to withstand, at

least to a far greater extent than the airship, the varying currents of the air.

The lifting principle must be analogous to that by which a kite is suspended. A kite is prevented from rising beyond a certain height by a string, and the pressure of the wind working against it at an angle tends to lift it, like a soft wedge continuously driven under it. In practice it makes no difference whether the kite be stationary in a wind or towed rapidly through a dead calm; the wedge-like action of the air remains the same.

Maxim decided upon constructing what was practically a huge compound kite driven by very powerful motors. But before setting to work on the machine itself he made some useful experiments to determine the necessary size of his kites or aeroplanes, and the force requisite to move them.

He accordingly built a 'whirling-table', consisting of a long arm mounted on a strong pivot at one end, and driven by a 10 horse-power engine. To the free end, which described a circle of 200 feet in circumference, he attached small aeroplanes, and by means of delicate balances discovered that at 40 miles an hour the aeroplane would lift 133 pounds per horse-power, and at 60 miles per hour every square foot of surface sustained 8 pounds weight. He, in common with other experimenters on the same lines, became aware of the fact that if it took a certain strain to suspend a stationary weight in the air, to advance it rapidly as well as to suspend it took a smaller strain. Now, as on sea and land, increased speed means a very rapid increase in the force required, this is a point in favour of the flying-machine. Professor Langley found that a brass plate weighing a pound, when whirled at great speed, was supported in the air by a pulling pressure of less than one ounce. And, of course, as the speed increased the plate became more nearly horizontal, offering less resistance to the air.

It is on this behaviour of the aeroplane that the hopes of Maxim and others have been based. The swiftly moving aeroplane, coming constantly on to fresh air, the inertia of which had not been disturbed, would resemble the skater who can at high speed traverse ice that would not bear him at rest.

Maxim next turned his attention to the construction of the aero-planes and engines. He made a special machine for testing fabrics, to decide which would be most suitable for stretching over strong frames to form the planes. The fabric must be light, very strong, and offer small frictional resistance to the air. The testing-machine was fitted with a nozzle, through which air was forced at a known pace on to the substance under trial, which met the air current at a certain angle and by means of indicators showed the strength of its 'lift' or tendency to rise, and that of its 'drift' or tendency to move horizontally in the direction of the air-current. A piece of tin, mounted at an angle of one in ten to the air current, showed a lift of ten times its drift. This proportion was made the standard. Experiments conducted on velvet, plush, silk, cotton and woollen goods proved that the drift of crape was several times that of its lift, but that fine linen had a lift equal to nine times its drift; while a sample of Spencer's balloon fabric was as good as tin.

Accordingly he selected this balloon fabric to stretch over light but strong frames. The stretching of the material was no easy matter, as uneven tension distorted it; but eventually the aeroplanes were completed, tight as drumheads. The large or central plane was 50 feet wide and 40 long; on either side were auxiliary planes, five pairs; giving a total area of 5,400 square feet.

The steam-engine built to give the motive power was perhaps the most interesting feature of the whole construction. Maxim employed steam in preference to any other power as being one with which he was most familiar, and yielding most force in proportion to the weight of the apparatus. He designed and constructed a pair of high-pressure compound engines, the high-pressure cylinders 5 inches in diameter, the low-pressure 8 inches, and both 1 foot stroke. Steam was supplied to the high-pressure cylinders at 320 pounds per square inch from a tubular boiler heated by a gasoline burner so powerful in its action as to raise the pressure from 100 to 200 pounds in a minute. The total weight of the boiler, burner, and engines developing 350 horse-power was 2,000 pounds, or about 6 pounds per horse-power. The two screw-propellers driven by the engine measured 17 feet 11 inches in diameter.

The completed flying-machine, weighing 7,500 pounds, was mounted on a railway-truck of 9-foot gauge, in Baldwyn's Park, Kent, not far from the gun, factories for which Sir Hiram is famous. Outside and parallel to the 9-foot track was a second track, 35 feet across, with a reversed rail, so that as soon as the machine should rise from the inner track long spars furnished with flanged wheels at their extremities should press against the under side of the outer track and prevent the machine from rising too far. Dynamometers, or instruments for measuring strains, were fitted to decide the driving and lifting power of the screws. Experiments proved that with the engines working at full power the screw-thrust against the air was 2,200 pounds, and the lifting force of the aeroplanes 10,000 pounds, or 1,500 in excess of the machine's weight.

Everything being ready the machine was fastened to a dynamometer and steam run up until it strained at its tether with maximum power; when the moorings were suddenly released and it bounded forward at a terrific pace, so suddenly that some of the crew were flung violently down on to the platform. When a speed of 42 miles was reached the inner wheels left their track, and the outer wheels came into play. Unfortunately, the long 35-foot axle trees were too weak to bear the strain, and one of them broke. The upper track gave way, and for the first time in the history of the world a flying-machine actually left the ground fully equipped with engines, boiler, fuel, and a crew. The journey, however, was a short one, for part of the broken track fouled the screws, snapped a propeller blade and necessitated the shutting off of the steam, which done, the machine settled to earth, the wheels sinking into the sward and showing by the absence of any marks that it had come directly downwards and not run along the surface.

The inventor was prevented by other business, and by the want of a sufficiently large open space, from continuing his experiments, which had demonstrated that a large machine heavier than air could be made to lift itself and move at high speed. Misfortune alone prevented its true capacities being shown.

Another experimenter on similar lines, but on a less heroic scale than Sir Hiram Maxim, is Professor S. P. Langley, the secretary of the Smithsonian Institution, Washington. For sixteen years he has devoted himself to a persevering course of study of the flying-machine, and after oft-repeated failures has scored a decided success in his Aerodrome, which, though only a model, has made considerable flights. His researches have proved beyond doubt that the amount of energy required for flight is but one-fiftieth of what was formerly regarded as a minimum. A French mathematician had proved by figures that a swallow must develop the power of a horse to maintain its rapid flight. Professor Langley's aerodrome has told a very different tale, affording another instance of the truth of the saying that an ounce of practice is worth a pound of theory.

A bird is nearly one thousand times heavier than the air it displaces. As a motor it develops huge power for its weight, and consumes a very large amount of fuel in doing so. An observant naturalist has calculated that the homely robin devours per diem, in proportion to its size, what would be to a man a sausage two hundred feet long and three inches thick! Any one who has watched birds pulling worms out of the garden lawn and swallowing them wholesale can readily credit this.

Professor Langley therefore concentrated himself on the production of an extremely light and at the same time powerful machine. Like Maxim, he turned to steam for motive-power, and by rigid economy of weight constructed an engine with boilers weighing 5 pounds, cylinders of 26 ounces, and an energy of 1 to 1.5 horsepower. Surely a masterpiece of mechanical workmanship. This he enclosed in a boat-shaped cover which hung from two pairs of aeroplanes 12.5 feet from tip to tip. The whole apparatus weighed nearly 30 pounds, of which one quarter represented the machinery. Experiments with smaller aerodromes warned the Professor that rigidity and balance were the two most difficult things to attain; also that the starting of the machine on its aerial course was far from an easy matter.

A soaring bird does not rise straight from the ground, but opens its wings and runs along the ground until the pressure of the air raises it sufficiently to give a full stroke of its pinions. Also it rises against the wind to get the full benefit of its lifting force. Professor Langley hired a houseboat on the Potomac River, and on the top of it built an apparatus from which the aerodrome could be launched into space at high velocity.

On May 6, 1896, after a long wait for propitious weather, the aerodrome was despatched on a trial trip. It rose in the face of the wind and travelled for over half a mile at the rate of twenty five miles an hour. The water and fuel being then exhausted it settled lightly on the water and was again launched. Its flight on both occasions was steady, and limited only by the rapid consumption of its power-producing elements. The Professor believes that larger machines would remain in the air for a long period and travel at speeds hitherto unknown to us.

The Aerocurve

In both the machines that we have considered the propulsive power was a screw. No counterpart of it is seen in Nature. This is not a valid argument against its employment, since no animal is furnished with driving-wheels, nor does any fish carry a revolving propeller in its tail. But some inventors are strongly in favour of copying Nature as regards the employment of wings. Mr. Sydney H. Hollands, an enthusiastic aeromobilist, has devised an ingenious cylinder-motor so arranged as to flap a pair of long wings, giving them a much stronger impulse on the down than on the up stroke. The pectoral muscles of a bird are reproduced by two strong springs which are extended by the upward motion of the wings and store up energy for the down-stroke. Close attention is also being paid to the actual shape of a bird's wing, which is not flat but hollow on its under side, and at the front has a slightly downward dip. 'Aerocurves' are therefore likely to supersede the 'aeroplane', for Nature would not have built bird's wings as they are without an object. The theory of the aerocurve's action is this: that the front of the wing on striking the air, gives it a downwards motion, and if the wing were quite flat its rear portion would strike air already in motion, and therefore less buoyant. The curvature of a floating bird's wings, which becomes more and more pronounced towards the rear, counteracts this yielding of the air by pressing harder upon it as it passes towards their hinder edge.

The aerocurve has been used by a very interesting group of experimenters, those who, putting motors entirely aside, have floated on wings, and learnt some of the secrets of balancing in the air. For a man to propel himself by flapping wings moved by legs or arms is impossible. Sir Hiram Maxim, in addressing the Aeronautical Society, once said that for a man to successfully imitate a bird his lungs must weigh 40 pounds, to consume sufficient oxygen, his breast muscles 75 pounds, and his breast bone be extended in front 21 inches. And unless his total weight were increased his legs must dwindle to the size of broomsticks, his head to that of an apple! So that for the present we shall be content to remain as we are!

Dr. Lilienthal, a German, was the first to try scientific wing-sailing. He became a regular air gymnast, running down the sides of an artificial mound until the wings lifted him up and enabled him to float a considerable distance before reaching earth again. His wings had an area of 160 square feet, or about a foot to every pound weight. He was killed by the wings collapsing in mid-air.

A similar fate also overtook Mr. Percy Pilcher, who abandoned the initial run down a sloping surface in favour of being towed on a rope attached to a fast moving vehicle. At present Mr. Octave Chanute, of Chicago, is the most distinguished member of the 'gliding' school. He employs, instead of wings, a species of kite made up of a number of small aerocurves placed one on the top of another a small distance apart. These box kites are said to give a great lifting force for their weight.

These and many other experimenters have had the same object in view-to learn the laws of equilibrium in the air. Until these are fully understood the construction of large flying-machines must be regarded as somewhat premature. Man must walk before he can run, and balance himself before he can fly.

There is no falling off in the number of aerial machines and schemes brought from time to time into public notice. We may assure ourselves that if patient work and experiment can do it the problem of 'how to fly' is not very far from solution at the present moment.

As a sign of the times, the War Office, not usually very ready to take up a new idea, has interested itself in the airship, and commissioned Dr. F. A. Barton to construct a dirigible balloon which combines the two systems of aerostation. Propulsion is effected by six sets of triple propellers, three on each side. Ascent is brought about partly by a balloon 180 feet long, containing 156,000 cubic feet of hydrogen, partly by nine aeroplanes having a total superficial area of nearly 2,000 square feet. The utilisation of these aeroplanes obviates the necessity to throw out ballast to rise, or to let out gas for a descent. The airship, being just heavier than air, is raised by the 135 horse-power motors pressing the aeroplanes against the air at the proper angle. In descent they act as parachutes.

The most original feature of this war balloon is the automatic water-balance. At each end of the 'deck' is a tank holding forty gallons of water. Two pumps circulate water through these tanks, the amount sent into a tank being regulated by a heavy pendulum which turns on the cock leading to the end which may be highest in proportion as it turns off that leading to the lower end. The idea is very ingenious, and should work successfully when the time of trial comes.

Valuable money prizes will be competed for by aeronauts at the coming World's Fair at St. Louis in 1903. Sir Hiram Maxim has expressed an intention of spending £20,000 in further experiments and prizes. In this country, too, certain journals have offered large rewards to any aeronaut who shall make prescribed journeys in a given time.

It has also been suggested that aeronautical research should be endowed by the state, since England has nothing to fear more than the flying machine and the submarine boat, each of which tends to rob her of the advantages of being an island by exposing her to unexpected and unseen attacks.

Tennyson, in a fine passage in 'Locksley Hall', turns a poetical eye towards the future. This is what he sees:

'For I dipt into the future, far as human eye could see,
Saw the vision of the world and all the wonder that would be,

Saw the heavens fill with commerce, argosies of magic sail,
Pilots of the purple twilight dropping down with costly bales,
Heard the heavens fill with shouting,
Then there rained a ghostly dew,
From the nations' airy navies,
Grappling in the central blue'.

Expressed in more prosaic language, the flying machine will primarily be used for military purposes. A country cannot spread a metal umbrella over itself to protect its towns from explosives dropped from the clouds.

Mail services will be revolutionised. The pleasure aerodrome will take the place of the yacht and motor-car, affording grand opportunities for the mountaineer and explorer (if the latter could find anything new to explore). Then there will also be a direct route to the North Pole over the top of those terrible ice-fields that have cost civilisation so many gallant lives. And possibly the ease of transit will bring the nations closer together, and produce good-fellowship and concord among them. It is pleasanter to regard the flying-machine of the future as a bringer of peace than as a novel means of spreading death and destruction.

Section Three

Visual Reproduction

Chapter 11

Animated Pictures

Has it ever occurred to the reader to ask himself why rain appears to fall in streaks though it arrives at earth in drops? Or why the glowing end of a charred stick produces fiery lines if waved about in the darkness? Common sense tells us the drop and the burning point cannot be in two places at one and the same time. And yet apparently we are able to see both in many positions simultaneously.

This seeming paradox is due to 'persistence of vision', a phenomenon that has attracted the notice of scientific men for many centuries. Persistence may be briefly explained thus:

The eye is extremely sensitive to light, and will, as is proved by the visibility of the electric spark, lasting for less than the millionth part of a second, receive impressions with marvellous rapidity. But it cannot get rid of these impressions at the same speed. The duration of a visual impression has been calculated as one-tenth to one-twenty-first of a second. The electric spark, therefore, appears to last much longer than it really does.

Hence it is obvious that if a series of impressions follow one another more rapidly than the eye can free itself of them, the impressions will overlap, and one of four results will follow:

Apparently uninterrupted presence of an image if the same image be repeatedly represented.

Confusion, if the images be all different and disconnected.

Combination, if the images, of two or a very few objects be presented in regular rotation.

Motion, if the objects be similar in all but one part, which occupies a slightly different portion in each presentation.

The Thaumatrope

In connection with the third point above, an interesting story is told of Sir J. Herschel by Charles Babbage. Quoted from Mr. Henry V. Hopwood's 'Living Pictures', to which book the author is indebted for much of his information in this chapter.

'One day Herschel, sitting with me after dinner, amusing himself by spinning a pear upon the table, suddenly asked whether I could show him the two sides of a shilling at the same moment. I took out of my pocket a shilling, and holding it up before the looking-glass, pointed out my method. 'No', said my friend, 'that won't do;' then spinning my shilling upon the table, he pointed out his method of seeing both sides at once. The next day I mentioned the anecdote to the late Dr. Fitton, who a few days after brought me a beautiful illustration of the principle. It consisted of a round disc of card suspended between two pieces of sewing silk. These threads being held between the finger and thumb of each hand, were then made to turn quickly, when the disc of card, of course, revolved also. Upon one side of this disc of card was painted a bird, upon the other side an empty bird-cage.

On turning the thread rapidly the bird appeared to have got inside the cage. We soon made numerous applications, as a rat on one side and a trap on the other, etc. It was shown to Captain Kater, Dr. Wollaston, and many of our friends, and was, after the lapse of a short time, forgotten. Some months after, during dinner at the Royal Society Club, Sir Joseph Banks being in the chair, I heard Mr. Barrow, then secretary to the Admiralty, talking very loudly about a wonderful invention of Dr. Paris, the object of which I could not quite understand. It was called the Thaumatrope, and was said to be sold at the Royal Institution, in Albemarle Street. Suspecting that it had some connection with our unnamed toy I went next morning and purchased for seven shillings and sixpence a thaumatrope, which I afterwards sent down to Slough to the late Lady Herschel. It was precisely the thing which her son and Dr. Fitton had contributed to invent, which amused all their friends for a time, and had then been forgotten.'

The thaumatrope, then, did nothing more than illustrate the power of the eye to weld together a couple of alternating impressions. The toys to which we shall next pass represent the same principle working in a different direction towards the production of the living picture.

Now, when we see a man running (to take an instance) we see the same body and the same legs continuously, but in different positions, which merge insensibly the one into the other. No method of reproducing that impression of motion is possible if only one drawing, diagram, or photograph be employed.

A man represented with as many legs as a centipede would not give us any impression of running or movement; and a blur showing the positions taken successively by his legs would be equally futile. Therefore we are driven back to a series of pictures, slightly different from one another; and in order that the pictures may not be blurred a screen must be interposed before the eye while the change from picture to picture is made. The shorter the period of change, and the greater the number of pictures presented to illustrate a single motion, the more realistic is the effect.

These are the general principles which have to be observed in all mechanism for the production of an illusory effect of motion. The persistence of vision has led to the invention of many optical toys, the names of which, in common with the names of most apparatus connected with the living picture, are remarkable for their length. Of these toys we will select three for special notice.

The Phenakistoscope

In 1833, Plateau of Ghent invented the phenakistoscope, 'the thing that gives one a false impression of reality', to interpret this formidable word. The phenakistoscope is a disc of card or metal round the edge of which are drawn a succession of pictures showing a man or animal in progressive positions. Between every two pictures a narrow slit is cut. The disc is mounted on an axle and revolved before a mirror, so that a person looking through the slits see one picture after another reflected in the mirror.

The Praxinoscope

The zoetrope, or Wheel of Life, which appeared first in 1860, is a modification of the same idea. In this instrument the pictures are arranged on the inner side of a hollow cylinder revolving on a vertical axis, its sides being perforated with slits above the pictures. As the slit in both cases caused distortion M. Reynaud, a Frenchman, produced in 1877 the praxinoscope, which differed from the zoetrope in that the pictures were not seen directly through slits, but were reflected by mirrors set half-way between the pictures and the axis of the cylinder, a mirror for every picture. Only at the moment when the mirror is at right angles to the line of sight would the picture be visible. M. Reynaud also devised a special lantern for projecting praxinoscope pictures on to a screen.

These and other somewhat similar contrivances, though ingenious, had very distinct limitations. They depended for their success upon the inventiveness and accuracy of the artist, who was confined in his choice of subject; and could, owing to the construction of the apparatus, only represent a small series of actions, indefinitely repeated by the machine. And as a complete action had to be crowded into a few pictures, the changes of position were necessarily abrupt.

To make the living picture a success two things were needed; some method of securing a very rapid series of many pictures, and a machine for reproducing the series, whatever its length. The method was found in, photography, with the advance of which the living picture's progress is so closely related, that it will be worth while to notice briefly the various improvements of photographic

processes. The old fashioned Daguerreotype process, discovered
in 1839, required an exposure of half-an-hour. The introduction of
wet collodion reduced this tax on a sitter's patience to ten seconds.
In 1878 the dry plate process had still further shortened the exposure to one second; and since that date the silver-salt emulsions
used in photography have had their sensitiveness to light so much
increased, that clear pictures can now be made in one-thousandth
of a second, a period minute enough to arrest the most rapid movements of animals.

By 1878, therefore, instantaneous photography was ready to aid
the living picture. Previously to that year series of photographs
had been taken from posed models, without however extending
the choice of subjects to any great extent. But between 1870 and
1880 two men, Marey and Muybridge, began work with the camera on the movements of horses. Marey endeavoured to produce
a series of pictures round the edge of one plate with a single lens
and repeated exposures. Muybridge, on the other hand, used a
series of cameras. He erected a long white back ground parallel to
which were stationed the cameras at equal distances. The shutters
of the cameras were connected to threads laid across the interval
between the background and the cameras in such a manner that a
horse driven along the track snapped them at regular intervals, and
brought about successive exposures. Muybridge's method was carried on by Anschutz, a German, who in 1899 brought out his electrical Tachyscope, or 'quick-seer'. Having secured his negatives
he printed off transparent positives on glass, and arranged these
last round the circumference of a large disc rotating in front of a
screen, having in it a hole the size of the transparencies. As each
picture came opposite the hole a Geissler tube was momentarily
lit up behind it by electrical contact, giving a fleeting view of one
phase of a horse's motion.

The introduction of the ribbon film in or about 1888 opened much
greater possibilities to the living picture than would ever have
existed had the glass plate been retained. It was now comparatively easy to take a long series of pictures; and accordingly we
find Messrs. Friese-Greene and Evans exhibiting in 1890 a camera

capable of securing three hundred exposures in half a minute, or ten per second.

The Kinetoscope

The next apparatus to be specially mentioned is Edison's Kinetoscope, which he first exhibited in England in 1894. As early as 1887 Mr. Edison had tried to produce animated pictures in a manner analogous to the making of a sound-record on a phonograph. He wrapped round a cylinder a sheet of sensitised celluloid which was covered, after numerous exposures, by a spiral line of tiny negatives. The positives made from these were illuminated in turn by flashes of electric light.

This method was, however, entirely abandoned in the perfected kinetoscope, an instrument for viewing pictures the size of a postage stamp, carried on a continuously moving celluloid film between the eye of the observer and a small electric lamp. The pictures passed the point of inspection at the rate of forty-six per second (a rate hitherto never approached), and as each picture was properly centred a slit in a rapidly revolving shutter made it visible for a very small fraction of a second. Holes punched at regular intervals along each side of the film engaged with studs on a wheel, and insured a regular motion of the pictures. This principle of a perforated film has been used by nearly all subsequent manufacturers of animatographs.

To secure forty-six negatives per second Edison invented a special exposure device. Each negative would have but one-forty-sixth of a second to itself, and that must include the time during which the fresh surface of film was being brought into position before the lens. He therefore introduced an intermittent gearing, which jerked the film forwards forty-six times per second, but allowed it to remain stationary for nine-tenths of the period allotted to each picture. During the time of movement the lens was covered by the shutter. This principle of exposure has also been largely adopted by other inventors. By its means weak negatives are avoided, while pictures projected on to a screen gain greatly in brilliancy and steadiness.

The capabilities of a long flexible film-band having been shown by Edison, he was not long without imitators. Phantoscopes, Bioscopes, Photoscopes, and many other instruments followed in quick succession. In 1895 Messrs. Lumiere scored a great success with their Cinematograph, which they exhibited at Marseilles and Paris; throwing the living picture as we now know it on to a screen for a large company to see. This camera lantern opens the era of commercial animated photography. The number of patents taken out since 1895 in connection with living-picture machines is sufficient proof that inventors have either found in this particular branch of photography a peculiar fascination, or have anticipated from it a substantial profit.

The Mutoscope and Biograph Company

A company known as the Mutoscope and Biograph Company has been formed for the sole object of working the manufacture and exhibition of the living picture on a great commercial scale. The present company is American, but there are subsidiary allied companies in many parts of the world, including the British Isles, France, Italy, Belgium, Germany, Austria, India, Australia, South Africa. The part that the company has played in the development of animated photography will be easily understood from the short account that follows.

The company controls three machines, the Mutograph, or camera for making negatives; the Biograph, or lantern for throwing pictures on to the screen; and the Mutoscope, a familiar apparatus in which the same pictures may be seen in a different fashion on the payment of a penny.

Externally the Mutograph is remarkable for its size, which makes it a giant of its kind. The complete apparatus weighs, with its accumulators, several hundreds of pounds. It takes a very large picture, as animatograph pictures go, two by two-and-a-half inches, which, besides giving increased detail, require less severe magnification than is usual with other films. The camera can make up to a hundred exposures per second, in which time twenty-two feet of film will have passed before the lens.

The film is so heavy that were it arrested bodily during each exposure and then jerked forward again, it might be injured. The mechanism of the mutograph, driven at regular speed by an electric motor, has been so arranged as to halt only that part of the film which is being exposed, the rest moving forward continuously. The exposed portion, together with the next surface, which has accumulated in a loop behind it, is dragged on by two rollers that are in contact with the film during part only of their revolutions. Thus the jerky motion is confined to but a few inches of the film, and even at the highest speeds the camera is peculiarly free from vibration.

An exposed mutograph film is wound for development round a skeleton reel, three feet in diameter and seven long, which rotates in a shallow trough containing the developing solution. Development complete, the reel is lifted from its supports and suspended over a succession of other troughs for washing, fixing, and final washing. When dry the negative film is passed through a special printing frame in contact with another film, which receives the positive image for the biograph. The difficulty of handling such films will be appreciated to a certain extent even by those whose experience is confined to the snaky behaviour of a short Kodak reel during development.

The Mutoscope Company's organisation is as perfect as its machinery. It has representatives in all parts of the world. Wherever stirring events are taking place, whether in peace or war, a mutograph operator will soon be on the spot with his heavy apparatus to secure pictures for world-wide exhibition. It need hardly be said that great obstacles, human and physical, have often to be overcome before a film can be exposed; and considerable personal danger encountered. We read that an operator, despatched to Cuba during the Spanish-American War was left three days and nights without food or water to guard his precious instruments, the party that had landed him having suddenly put to sea on sighting a Spanish cruiser. Another is reported to have had a narrow escape from being captured at sea by the Spaniards after a hot chase. It is also on record that a mutograph set up in Atlantic City to take a procession of fire-engines was charged and shattered by one of the engines; that the operators were flung into the crowd: and that nevertheless the box containing the exposed films was uninjured, and on development yielded a very sensational series of pictures lasting to the moment of collision.

The Mutoscope Company owns several thousand series of views, none probably more valuable than those of his Holiness the Pope, who graciously gave Mr. W. K. Dickson five special sittings, during which no less than 17,000 negatives were made, each one of great interest to millions of people throughout the world.

The company spares neither time nor money in its endeavour to supply the public with what will prove acceptable. A year's output runs into a couple of hundred miles of film. As much as 700 feet is sometimes expended on a single series, which may be worth anything up to £1,000.

The energy displayed by the operators is often marvellous. To take instances. The Derby of 1898 was run at 3.20 P.M. At ten o'clock the race was run again by Biograph on the great sheet at the Palace Theatre. On the home-coming of Lord Kitchener from the Soudan Campaign, a series of photographs was taken at Dover in the afternoon and exhibited the same evening! Or again, to consider a wider sphere of action, the Jubilee Procession of 1897 was watched in New York ten days after the event; two days later in Chicago; and in three more the films were attracting large audiences in San Francisco, 5,000 miles from the actual scene of the procession!

One may easily weary of a series of single views passed slowly through a magic-lantern at a lecture or entertainment. But when the Biograph is flashing its records at lightning speed there is no cause for dulness. It is impossible to escape from the fascination of movement. A single photograph gives the impression of mere resemblance to the original; but a series, each reinforcing the signification of the last, breathes life into the dead image, and deludes us into the belief that we see, not the representation of a thing, but the thing itself. The bill of fare provided by the Biograph Company is varied enough to suit the most fastidious taste. Now it is the great Naval Review off Spithead, or President Faure shooting pheasants on his preserves near Paris. A moment's pause and then the magnificent Falls of Niagara foam across the sheet; Maxim guns fire harmlessly; panoramic scenes taken from locomotives running at high velocity unfold themselves to the delighted spectators, who feel as if they really were speeding over open country, among towering rocks, or plunging into the darkness of a tunnel. Here is an express approaching with all the quiver and fuss of real motion, so faithfully rendered that it seems as if a catastrophe were imminent; when, snap! We are transported a hundred miles to watch it glide into a station. The doors open, passengers step out and shake hands with friends, porters bustle about after luggage, doors are slammed again, the guard waves his flag, and the carriages move slowly out of the picture. Then our attention is switched away to the 10-inch disappearing gun, landing and firing at Sandy Hook. And next, as though to show that nothing is beneath the notice of the biograph, we are perhaps introduced to a family of small pigs feeding from a trough with porcine earnestness and want of manners.

It must not be thought that the Living Picture caters for mere entertainment only. It serves some very practical and useful ends. By its aid the movements of machinery and the human muscles may be studied in detail, to aid a mechanical or medical education. It furnishes art schools with all the poses of a living model. Less serious pursuits, such as dancing, boxing, wrestling and all athletic sports and exercise, will find a use for it. As an advertising medium it stands unrivalled, and we shall owe it a deep debt of gratitude if it ultimately supplants the flaring posters that disfigure our towns and desecrate our landscapes. Not so long since, the directors of the Norddeutscher Lloyd Steamship Company hired the biograph at the Palace Theatre, London, to demonstrate to anybody who cared to witness a very interesting exhibition that their line of vessels should always be used for a journey between England and America.

The Living Picture has even been impressed into the service of the British Empire to promote emigration to the Colonies. Three years ago Mr. Freer exhibited at the Imperial Institute and in other places in England a series of films representing the 1897 harvest in Manitoba. Would-be emigrants were able to satisfy themselves that the great Canadian plains were fruitful not only on paper. For could they not see with their own eyes the stately procession of automatic binders reaping, binding, and delivering sheaves of wheat, and puffing engines threshing out the grain ready for market? A far preferable method this to the bogus descriptions of land companies such as lured poor Chuzzlewit and Mark Tapley into the deadly swamps of 'Eden'.

Again, what more calculated to recruit boys for our warships than the fine Polytechnic exhibition known as 'Our Navy'? What words, spoken or printed, could have the effect of a series of vivid scenes truthfully rendered, of drills on board ship, the manning and firing of big guns, the limbering-up of smaller guns, the discharge of torpedoes, the headlong rush of the 'destroyers'?

The Mutoscope, to which reference has been made above, may be found in most places of public entertainment, in refreshment bars, on piers, in exhibitions, on promenades. A penny dropped into a slot releases a handle, the turning of which brings a series of pictures under inspection. The pictures, enlarged from mutograph films, are mounted in consecutive order round a cylinder, standing out like the leaves of a book., When the cylinder, is revolved by means of the handle the picture cards are snapped past the eye, giving an effect similar to the lifelike projections on a biograph screen. From 900 to 1,000 pictures are mounted on a cylinder.

The advantages of the mutoscope; its convenient size, its simplicity, and the ease with which its contents may be changed to illustrate the topics and events of the day, have made the animated photograph extremely popular. It does for vision what the phonograph does for sound. In a short time we shall doubtless be provided with handy machines combining the two functions and giving us double value for our penny.

The real importance and value of animated photography will be more easily estimated a few years hence than to-day, when it is still more or less of a novelty. The multiplication of illustrated newspapers and magazines points to a general desire for pictorial matter to help down the daily, weekly, or monthly budget of news, even if the illustrations be imaginative products of Fleet Street rather than faithful to fact. The reliable living picture (we except the 'set-scene') which 'holds up a mirror to nature', will be a companion rather than a rival of journalism, following hard on the description in print of an event that has taken place under the eye of the recording camera.

The zest with which we have watched during the last two years biographic views of the embarkation and disembarkation of troops, of the transport of big guns through drifts and difficult country, and of the other circumstances of war, is largely due to the descriptions we have already read of the things that we see on the screen. And, on the other hand, the impression left by a series of animated views will dwell in our memories long after the contents of the newspaper columns have become confused and jumbled. It is therefore especially to be hoped that photographic records will be kept of historic events, such as the Jubilee, the Queen's Funeral, King Edward's Coronation, so that future generations may, by the turning of a handle, be brought face to face with the great doings of a bygone age.

Chapter 12

Photography in Colours

While photography was still in its infancy many people believed that, a means having been found of impressing the representation of an object on a sensitised surface, a short time only would have to elapse before the discovery of some method of registering the colours as well as the forms of nature.

Photography has, during the last forty years, passed through some startling developments, especially as regards speed. Experts, such as M. Marey, have proved the superiority of the camera over the human eye in its power to grasp the various phases of animal motion. Even rifle bullets have been arrested in their lightning flight by the sensitised plate.

But while the camera is a valuable aid to the eye in the matter of form, the eye still has the advantage so far as colour is concerned. It is still impossible for a photographer by a simple process similar to that of making an ordinary black-and-white negative, to affect a plate in such a manner that from it prints may be made by a single operation showing objects in their natural colours. Nor, for the matter of that, does colour photography direct from nature seem any nearer attainment now than it was in the time of Daguerre.

There are, however, extant several methods of making colour photographs in an indirect or roundabout way. These various 'dodges' are, apart from their beautiful results, so extremely ingenious and interesting that we propose to here examine three of the best known.

The reader must be careful to banish from his mind those coloured photographs so often to be seen in railway carriages and shop windows, which are purely the result of hand-work and mechanical printing, and therefore not colour photographs at all.

Before embarking on an explanation of these three methods it will be necessary to examine briefly the nature of those phenomena on which all are based; light and colour. The two are really identical, light is colour and colour is light.

Scientists now agree that the sensation of light arises from the wave-like movements of that mysterious fluid, the omnipresent ether. In a beam of white light several rates of wave vibrations exist side by side. Pass the beam through a prism and the various rapidities are sorted out into violet, indigo, blue, green, yellow, orange and red, which are called the pure colours, since if any of them be passed again through a prism the result is still that colour. Crimson, brown, etc., the composite colours, would, if subjected to the prism, at once split up into their component pure colours.

There are several points to be noticed about the relationship of the seven pure colours. In the first place, though they are all allies in the task of making white light, there is hostility among them, each being jealous of the others, and only waiting a chance to show it. Thus, suppose that we have on a strip of paper squares of the seven colours, and look at the strip through a piece of red glass we see only one square, the red, in its natural colour, since that square is in harmony only with red rays. (Compare the sympathy of a piano with a note struck on another instrument; if C is struck, say on a violin, the piano strings producing the corresponding note will sound, but the other strings will be silent.) The orange square suggests orange, but the green and blue and violet appear black. Red glass has arrested their ether vibrations and said 'no way here'. Green and violet would serve just the same trick on red or on each other. It is from this readiness to absorb or stop dissimilar rays that we have the different colours in, a landscape flooded by a common white sunlight. The trees and grass absorb all but the green rays, which they reflect. The dandelions and buttercups capture and hold fast all but the yellow rays. The poppies in the corn send us back red only, and the cornflowers only blue; but the daisy is more generous and gives up all the seven. Colour therefore is not a thing that can be touched, any more than sound, but merely the capacity to affect the retina of the eye with a certain number of ether vibrations per second, and it makes no difference whether light is reflected from a substance or refracted through a substance; a red brick and a piece of red glass have similar effects on the eye.

This then is the first thing to be clearly grasped, that whenever a colour has a chance to make prisoners of other colours it will do so.

The second point is rather more intricate, viz. that this imprisonment is going on even when friendly concord appears to be the order of the day. Let us endeavour to present this clearly to the reader. Of the pure colours, violet, green and red; the extremes and the centre, are sufficient to produce white, because each contains an element of its neighbours. Violet has a certain amount of indigo, green some yellow, red some orange; in fact every colour of the spectrum contains a greater or less degree of several of the others, but not enough to destroy its own identity. Now, suppose that we have three lanterns projecting their rays on to the same portion of a white sheet, and that in front of the first is placed a violet glass, in front of the second a green glass, in front of the third a red glass. What is the result? A white light. Why? Because they meet on equal terms, and as no one of them is in a point of advantage no prisoners can be made and they must work in harmony. Next, turn down the violet lantern, and green and red produce a yellow, halfway between them; turn down red and turn up violet, indigo blue results. All the way through a compromise is effected.

But supposing that the red and green glasses are put in front of the same lantern and the white light sent through them-where has the yellow gone to? only a brownish-black light reaches the screen. The same thing happens with red and violet or green and violet.

Prisoners have been taken, because one colour has had to demand passage from the other. Red says to green, 'You want your rays to pass through me, but they shall not'. Green retorts, 'Very well; but I myself have already cut off all but green rays, and if they don't pass you, nothing shall'. And the consequence of the quarrel is practical darkness.

The same phenomenon may be illustrated with blue and yellow. Lights of these two colours projected simultaneously on to a sheet yield white; but white light sent through blue and yellow glass in succession produces a green light. Also, blue paint mixed with yellow gives green. In neither case is there darkness or entire cutting-off of colour, as in the case of Red + Violet or Green + Red. The reason is easy to see.

Blue light is a compromise of violet and green; yellow of green and red. Hence the two coloured lights falling on the screen make a combination which can be expressed as an addition sum.

Blue = green +violet

Yellow = green + red

Green + violet + red = white

But when light is passed through two coloured glasses in succession, or reflected from two layers of coloured paints, there are prisoners to be made.

Blue passes green and violet only.

Yellow passes green and red only.

So violet is captured by yellow, and red by blue, green being free to pass on its way.

There is, then, a great difference between the mixing of colours, which evokes any tendency to antagonism, and the adding of colours under such conditions that they meet on equal terms. The first process happens, as we have seen, when a ray of light is passed through colours in succession; the second, when lights stream simultaneously on to an object. A white screen, being capable of reflecting any colour that falls on to it, will with equal readiness show green, red, violet, or a combination; but a substance that is in white light red, or green, or violet will capture any other colour. So that if for the white screen we substituted a red one, violet or green falling simultaneously, would yield blackness, because red takes both prisoners; if it were violet, green would be captured, and so on.

From this follows another phenomenon: that whereas projection of two or more lights may yield white, white cannot result from any mixture of pigments. A person with a whole box full of paints could not get white were he to mix them in an infinitude of different ways; but with the aid of his lanterns and as many differently coloured glasses the feat is easy enough.

Any two colours which meet on equal terms to make white are called complementary colours.

Thus yellow (= red + green lights) is complementary of violet.

Thus pink (= red + violet lights) is complementary of green.

Thus blue (=violet + green lights) is complementary of red.

This does not of course apply to mixture of paints, for complementary colours must act together, not in antagonism.

If the reader has mastered these preliminary considerations he will have no difficulty in following out the following processes.

The Joly Process

Invented by Professor Joly of Dublin. A glass plate is ruled across with fine parallel lines; 350 to the inch, we believe. These lines are filled in alternately with violet, green, and red matter, every third being violet, green or red as the case may be. The colour-screen is placed in the camera in front of the sensitised plate. Upon an exposure being made, all light reflected from a red object (to select a colour) is allowed to pass through the red lines, but blocked by all the green and violet lines. So that on development that part of the negative corresponding to the position of the red object will be covered with dark lines separated by transparent belts of twice the breadth. From the negative a positive is printed, which of course shows transparent lines separated by opaque belts of twice their breadth. Now, suppose that we take the colour screen and place it again in front of the plate in the position it occupied when the negative was taken, the red lines being opposite the transparent parts of the positive will be visible, but the green and violet being blocked by the black deposit behind them will not be noticeable. So that the object is represented by a number of red lines, which at a small distance appear to blend into a continuous whole.

The violet and green affect the plate in a corresponding manner; and composite colours will affect two sets of lines in varying degrees, the lights from the two sets blending in the eye. Thus yellow will obtain passage from both green and red, and when the screen is held up against the positive, the light streaming through the green and red lines will blend into yellow in the same manner as they would make yellow if projected by lanterns on to a screen. The same applies to all the colours.

The advantage of the Joly process is that in it only one negative has to be made.

The Ives Process

Mr. Frederic Eugene Ives, of Philadelphia, arrives at the same result as Professor Joly, but by an entirely different means. He takes three negatives of the same object, one through a violet-blue, another through a green, and a third through a red screen placed in front of the lens. The red negative is affected by red rays only; the green by green rays only, and the violet-blue by violet-blue rays only, in the proper gradations. That is to say, each negative will have opaque patches wherever the rays of a certain kind strike it; and the positive printed off will be by consequence transparent at the same places. By holding the positive made from the red-screen negative against a piece of red glass, we should see light only in those parts of the positive which were transparent. Similarly with the green and violet positives if viewed through glasses of proper colour. The most ingenious part of Mr. Ives' method is the apparatus for presenting all three positives (lighted through their coloured glasses) to the eye simultaneously. When properly adjusted, so that their various parts exactly coincide, the eye blends the three together, seeing green, red, or violet separately, or blended in correct proportions. The Kromoscope, as the viewing apparatus is termed, contains three mirrors, projecting the reflections from the positives in a single line. As the three slides are taken stereoscopically the result gives the impression of solidity as well as of colour, and is most realistic.

The Sanger Shepherd Process

This is employed mostly for lantern transparencies. As in the Ives process, three negatives and three transparent positives are made. But instead of coloured glasses being used to give effect to the positives the positives themselves are dyed, and placed one on the top of another in close contact, so that the light from the lantern passes through them in succession. We have therefore now quitted the realms of harmony for that of discord, in which prisoners are made; and Mr. Shepherd has had to so arrange matters that in every case the capture of prisoners does not interfere with the final result, but conduces to it.

In the first place, three negatives are secured through violet, green, and red screens. Positives are printed by the carbon process on thin celluloid films. The carbon film contains gelatine and bichromate of potassium. The light acts on the bichromate in such a way as to render the gelatine insoluble. The result is that, though in the positives there is at first no colour, patches of gelatine are left which will absorb dyes of various colours. The dyeing process requires a large amount of care and patience.

Now, it would be a mistake to suppose that each positive is dyed in the colour of the screen through which its negative was taken. A moment's consideration will show us why.

Let us assume that we are photographing a red object, a flower-pot for instance. The red negative represents the pot by a dark deposit. The positive printed off will consequently show clear glass at that spot, the unaffected gelatine being soluble. So that to dye the plate would be to make all red except the very part which we require red; and on holding it up to the light the flower-pot would appear as a white transparent patch. How then is the problem to be solved?

Mr. Shepherd's process is based upon an ordered system of prisoner-taking. Thus, as red in this particular case is wanted it will be attained by the other two positives (which are placed in contact with the red positive, so that all three coincide exactly), robbing white light of all but its red rays.

Now, if the other positives were dyed green and violet, what would happen? They would not produce red, but by robbing white light between them of red, green, and violet, would produce blackness, and we should be as far as ever from our object.

The positives are therefore dyed, not in the same colours as the screens used when the negatives were made, but in their complementary colours, i.e. as explained above, those colours which added to the colour of the screen would make white.

The red screen negative is dyed (violet + green) = blue.

The green negative (red + violet) = pink.

The violet negative (red + green) = yellow.

To return to our flower-pot. The red-screen positive (dyed blue) is, as we saw, quite transparent where the pot should be. But behind the transparent gap are the pink and yellow positives.

White light (= violet + green + red) passes through pink (= violet + red), and has to surrender all its green rays.

The violet and red pass on and encounter yellow (= green + red), and violet falls a victim to green, leaving red unmolested.

If the flower-pot had been white all three positives would have contained clear patches unaffected by the three dyes, and the white light would have been unobstructed. The gradations and mixtures of colours are obtained by two of the screens being influenced by the colour of the object. Thus, if it were crimson, both violet and red-screen negatives would be affected by the rays reflected by it, and the green screen negative not at all. Hence the pink positive would be pink, the yellow clear, and the blue clear.

White light passing through is robbed by pink of green, leaving red + violet = crimson.

Colour Printing

Printing in ink colours is done in a manner very similar to the Sanger Shepherd lantern slide process. Three blocks are made, by the help of photography, through violet, green and red screens, and etched away with acid, like ordinary half-tone black-and white blocks. The three blocks have applied to them ink of a complementary colour to the screen they represent, just as in the Sanger Shepherd process the positives were dyed. The three inks are laid over one another on the paper by the blocks, the relieved parts of which (corresponding to the undissolved gelatine of the Shepherd positives) only take the ink. White light being reflected through layers of coloured inks is treated in just the same way as it would be were it transmitted through coloured glasses, yielding all the colours in approximately correct gradations.

Chapter 13

Type-setting by Machinery

To the Assyrian brick makers who, thousands of years ago, used blocks wherewith to impress on their unbaked bricks hieroglyphics and symbolical characters, must be attributed the first hesitating step towards that most marvellous and revolutionary of human discoveries; the art of printing. Not, however, till the early part of the fifteenth century did Gutenberg and Coster conceive the brilliant but simple idea of printing from separate types, which could be set in different orders and combinations to represent different ideas. For Englishmen, 1474 deserves to rank with 1815, as in that year a very Waterloo was won on English soil against the forces of ignorance and oppression, though the effects of the victory were not at once evident.

Considering the stir made at the time by the appearance of Caxton's first book at Westminster, it seems strange that an invention of such importance as the printing-press should have been frowned upon by those in power, and so discouraged that for nearly two centuries printing remained an ill-used and unprogressive art, a giant half strangled in his cradle. Yet as soon as prejudice gave it an open field, improved methods followed close on one another's heels. To-day we have, in the place of Caxton's rude hand-made press, great cylinder machines capable of absorbing paper by the mile, and grinding out 20,000 impressions an hour as easily as a child can unwind a reel of cotton.

Side by side with the problem how to produce the greatest possible number of copies in a given time from one machine, has arisen another; how to set up type with a proportionate rapidity. A press without type is as useless as a chaff-cutter without hay or straw. The type once assembled, as many casts or stereotypes can be made from it as there are machines to be worked. But to arrange a large body of type in a short time brings the printer face to face with the need of employing the expensive services of a small army of compositors, unless he can attain his end by some equally efficient and less costly means. For the last century a struggle has been in progress between the machine compositor and the human compositor, mechanical ingenuity against eye and brains. In the last five years the battle has turned most decidedly in favour of the machine. To-day there are in existence two wonderful contrivances which enable a man to set up type six times as fast as he could by hand from a box of type, with an ease that reminds one of the mythical machine for the conversion of live pigs into strings of sausages by an uninterrupted series of movements.

These machines are called respectively the Linotype and Monotype. Roughly described, they are to the compositor what a typewriter is to a clerk, forming words in obedience to the depression of keys on a keyboard. But whereas the typewriter merely imprints a single character on paper, the Linotype and Monotype cast, deliver, and set up type from which an indefinite number of impressions can be taken. They meet the compositor more than half-way, and simplify his labour while hugely increasing his productiveness.

The Linotype

As far back as 1842 periodicals were mechanically composed by a machine which is now practically forgotten. Since that time hundreds of other inventions have been patented, and some scores of different machines tried, though with small success in most cases; as it was found that quality of composition was sacrificed to quantity, and that what at first appeared a short cut to the printing-press was after all the longest way round, when corrections had all been attended to. A really economical type-setter must be accurate as well as prolific. Slipshod work will not pay in the long run.

Such a machine was perfected a few years ago by Ottmar Mergenthaler of Baltimore, who devised the plan of casting a whole line of type. The Linotype Composing Machine, to give it its full title, produces type all ready for the presses in 'slugs' or lines, hence the name, Lin-o-type. It deserves at least a short description.

The Linotype occupies about six square feet of floor space, weighs one ton, and is entirely operated by one man. Its most prominent features are a sloping magazine at the top to hold the brass matrices, or dies from which the type is cast, a keyboard controlling the machinery to drop and collect the dies, and a long lever which restores the dies to the magazine when done with.

The operator sits facing the keyboard, in which are ninety keys, variously coloured to distinguish the different kinds of letters. His hands twinkle over the keys, and the brass dies fly into place. When a key is depressed a die shoots from the magazine on to a travelling belt and is whirled off to the assembling-box. Each die is a flat, oblong brass plate, of a thickness varying with the letter, having a large V-shaped notch in the top, and the letter cut half-way down on one of the longer sides. A corresponding letter is stamped on the side nearest to the operator so that he may see what he is doing and make needful corrections.

As soon as a word is complete, he touches the 'spacing' lever at the side of the keyboard. The action causes a 'space' to be placed against the last die to separate it from the following word. The operations are repeated until the tinkle of a bell warns him that, though there may be room for one or two more letters, the line will not admit another whole syllable. The line must therefore be 'justified', that is, the spaces between the words increased till the vacant room is filled in. In hand composition this takes a considerable time, and is irksome; but at the linotype the operator merely twists a handle and the wedge-shaped 'spaces', placed thin end upwards, are driven up simultaneously, giving the lateral expansion required to make the line of the right measure.

A word about the 'spaces', or space-bands. Were each a single wedge the pressure would be on the bottom only of the dies, and their tops, being able to move slightly, would admit lead between them. To obviate this a small second wedge, thin end downwards, is arranged to slide on the larger wedge, so that in all positions parallelism is secured. This smaller wedge is of the same shape as the dies and remains stationary in line with them, the larger one only moving.

The line of dies being now complete, it is automatically borne off and pressed into contact with the casting wheel. This wheel, revolving on its centre, has a slit in it corresponding in length and width to the size of line required. At first the slit is horizontal, and the dies fit against it so that the row of sunk letters on the faces are in the exact position to receive the moulten lead, which is squirted through the slit from behind by an automatic pump, supplied from a metal-pot. The pot is kept at a proper heat of 550° Fahrenheit by the flames of a Bunsen burner.

The lead solidifies in an instant, and the 'slug' of type is ready for removal, after its back has been carefully trimmed by a knife. The wheel revolves for a quarter-turn, bringing the slit into a vertical position; a punch drives out the 'slug', which is slid into the galley to join its predecessors. The wheel then resumes its former horizontal position in readiness for another cast.

The assembled dies have for the time done their work and must be returned to the magazine. The mechanism used to effect this is peculiarly ingenious.

An arm carrying a ribbed bar descends. The dies are pushed up, leaving the 'spaces' behind to be restored to their proper compartment, till on a level with the ribbed bar, on to which they are slid by a lateral movement, the notches of the V-shaped opening in the top side of each die engaging with the ribs on the bar. The bar, then ascends till it is in line with a longer bar of like section passing over the open top of the entire magazine. A set of horizontal screw-bars, rotating at high speed, transfer the dies from the short to the long bar, along which they move till, as a die comes above its proper division of the magazine, the arrangement of the teeth allows it to drop. While all this has been going on, the operator has composed another line of moulds, which will in turn be transferred to the casting wheel, and then back to the magazine. So that the three operations of composing, casting, and sorting moulds are in progress simultaneously in different parts of the machine; with the result that as many as 20,000 letters can be formed by an expert in the space of an hour, against the 1,500 letters of a skilled hand compositor.

How about corrections? Even a comma too few or too many needs the whole line cast over again. It is a convincing proof of the difference in speed between the two methods that a column of type can be corrected much faster by the machine, handicapped as it is by its solid 'slugs', than by hand. No wonder then that more than 1,000 Linotypes are to be found in the printing offices of Great Britain.

The Monotype

The Monotype, like the Linotype, aims at speed in composition, but in its mechanism it differs essentially from the Linotype. In the first place, the apparatus is constructed in two quite separate parts. There is a keyboard, which may be on the third floor of the printing offices, and the casting machine, which ceaselessly casts and sets type in the basement. Yet they are but one whole. The connecting link is the long strip of paper punched by the keyboard mechanism, and then transferred to the casting machine to bring about the formation of type. The keyboard is the servant of man; the casting machine is the slave of the keyboard.

Secondly, the Monotype casts type, not in blocks or a whole line, but in separate letters. It is thus a complete type-foundry. Order it to cast G's and it will turn them out by the thousand till another letter is required.

Thirdly, by means of the punched paper roll, the same type can be set up time after time without a second recourse to the keyboard, just as a tune is ground repeatedly out of a barrel organ.

The keyboard has a formidable appearance. It contains 225 keys, providing as many characters; also thirty keys to regulate the spacing of the words. At the back of the machine a roll of paper runs over rollers and above a row of thirty little punches worked by the keys. A key being depressed, an opened valve admits air into two cylinders, each driving a punch. The punches fly up and cut two neat little holes in the paper. The roll then moves forward for the next letter. At the end of the word a special lever is used to register a space, and so on to the end of the line. The operator then consults an automatic indicator which tells him exactly how much space is left, and how much too long or too short the line would be if the spaces were of the normal size. Supposing, for instance, that there are ten spaces, and that there is one-tenth of an inch to spare. It is obvious that by extending each space one-hundredth of an inch the vacant room will be exactly filled. Similarly, if the ten normal spaces would make the line one-tenth of an inch too long, by decreasing the spaces each one-hundredth inch the line will also be 'justified'.

But the operator need not trouble his head about calculations of this kind. His indicator, a vertical cylinder covered with tiny squares, in each of which are printed two figures, tell him exactly what he has to do. On pressing a certain key the cylinder revolves and comes to rest with the tip of a pointer over a square. The operator at once presses down the keys bearing the numbers printed on that square, confident that the line will be of the proper length.

As soon as the roll is finished, it is detached from the keyboard and introduced to the casting machine. Hitherto passive, it now becomes active. Having been placed in position on the rollers it is slowly unwound by the machinery. The paper passes over a hollow bar in which there are as many holes as there were punches in the keyboard, and in precisely the same position. When a hole in the paper comes over a hole in the hollow bar air rushes in, and passing through a tube actuates the type-setting machinery in a certain manner, so as to bring the desired die into contact with moulten lead. The dies are, in the Monotype, all carried in a magazine about three inches square, which moves backwards or forwards, to right or left, in obedience to orders from the perforated roll. The dies are arranged in exactly the same way as the keys on the keyboard. So that, supposing A to have been stamped on the roll, one of the perforations causes the magazine to slide one way, while the other shoves it another, until the combined motions bring the matrix engraved with the A underneath the small hole through which moulten lead is forced. The letter is ejected and moves sideways through a narrow channel, pushing preceding letters before it, and the magazine is free for other movements.

At the end of each word a 'space' or blank lead is cast, its size exactly determined by the justifying hole belonging to that line. Word follows word till the line is complete; then a knife-like lever rises, and the type is propelled into the 'galley'. Though a slave the casting machine will not tolerate injustice. Should the compositor have made a mistake, so that the line is too long or too short, automatic machinery at once comes into play, and slips the driving belt from the fixed to the loose pulley, thus stopping the machine till someone can attend to it. But if the punching has been correctly done, the machine will work away unattended till, a whole column of type having been set up, it comes to a standstill.

The advantages of the Monotype are easily seen. In order to save money a man need not possess the complete apparatus. If he has the keyboard only he becomes to a certain extent his own compositor, able to set up the type, as it were by proxy, at any convenient time. He can give his undivided attention to the keyboard, stop work whenever he likes without keeping a casting-machine idle, and as soon as his roll is complete forward it to a central establishment where type is set. There a single man can superintend the completion of half-a-dozen men's labours at the keyboard. That means a great reduction of expense.

In due time he receives back his copy in the shape of set-up type, all ready to be corrected and transferred to the printing machines. The type done with, he can melt it down without fear of future regret, for he knows that the paper roll locked up in his cupboard will do its work a second time as well as it did the first. Should he need the same matter re-setting, he has only to send the roll through the post to the central establishment.

Thanks to Mr. Lanston's invention we may hope for the day when every parish will be able to do its own printing, or at least set up its own magazine. The only thing needful will be a Monotype key-board supplied by an enlightened Parish Council, as soon as the expense appears justifiable, and kept in the Post Office or Village Institute. The payment of a small fee will entitle the Squire to punch out his speech on behalf of the Conservative Candidate, the Schoolmaster to compose special information for his pupils, the Rector to reduce to print pamphlets and appeals to charity. And if those of humbler degree think they can strike eloquence from the keys, they too will of course be allowed to turn out their ideas literally by the yard.

Book Production

The manuscript was scanned in from an early printed edition of the book and OCR (optical character recognition) was performed on the resulting images to give a text file. The text was then corrected and imported into Sun Star Office Writer; a word processing program distributed by Sun and compliant with the Open Office standard. After being edited in Star Office Writer it was exported in the DocBook format, which has a simple XML structure. This XML file was imported into Adobe InDesign using a custom written XSLT file to restructure it. After typesetting and proofing, the manuscript was exported to PDF and uploaded to the printers.

Typeset in Times New Roman, with headings and titles set in Perpetua with an 80% tint.

Also by Bosko Books

Bosko Books publishes books about people and technology. This includes design for use, digital media, and information technology; current, future and historical. You can order the following books on the publishers web site: www.boskobooks.com or on Amazon by searching for the ISBN number.

Designing the Real World

by Lon Barfield
ISBN 0954723910

One of the most popular columns in the SIGCHI bulletin (the ACM magazine for interaction designers) was Barfield's Real World column, observing everyday interactions in both real and digital environments. This book contains fifty of those columns, covering such fundamental topics as:

Switching things on and off

Choosing the correct terminology for interfaces

Designing volume controls

Annoying sounds coming from alarm-clocks

Making the ideal slice of toast

They have been gathered together, along with extra sections on observing the real world and a number of new columns. The book is both entertaining and enlightening. Lecturers will find it a good supplement to any course dealing with designing for people, while industrial and digital designers can learn from the observations and insights it contains. The content is an inspirational resource for interaction designers, web designers, architects, industrial designers and anybody who has ever said 'Who on earth designed that?!'

The User Interface: Concepts and Design

by Lon Barfield
ISBN 0954723902

Everybody has problems using technology, from heating controls through to TV program guides. Move to computers and the problems are even worse; even the simplest digital systems seem to behave in strange ways. This book considers the problems of usability of technology and examines the factors that play a role in the design of such systems. Its goal is to introduce students and those working in related areas to the issues and to support them in analyzing problems and coming up with their own designs. It covers the issues surrounding the design of everyday technology before bringing computers into the picture and looking at how those issues change with the design of the user interface to computer systems. There are plenty of good seminar style exercises with accompanying guidelines.

The text uses numerous real-world examples to get its message across and it does so in an amusing and authoritative style. It steers clear of technical issues which means that it is very general in nature, that it retains its relevance as technologies change and that the text does not get bogged down in technical jargon. As well as the exercises, each chapter has an imaginary dialogue between Hemelsworth; a frustrated lord, and his dim-witted butler Barker, who is prone to behaving like your average computer system.

First printed and reprinted by Addison Wesley, this timeless title is now available from Bosko Books. It is still relevant and useful, and continues to be used to teach interaction design courses and computing courses relating to the user interface.

Dust or Magic

by Bob Hughes
ISBN 0954723953

Dust or Magic was primarily written for the young, talented people whose creative instincts are kindled by computers and live to create 'good stuff', but who are systematically betrayed by the managerial types in suits who hire them, set them absurd tasks, and sack them when their half-baked schemes go belly-up. It is also for people who simply want to know how human creativity fares in the digital age.

Originally published by Addison-Wesley (under the title 'Dust or Magic, Secrets of successful multimedia design') this book is, in part, a 'secret history' of computers: a history told from the vantage point of the people who did the work. We have insiders' accounts of a range of influential products and projects, many of which were in danger of being forgotten. The scene is illuminated by recent insights into creativity and well-being from the fields of psychology and neuroscience, as well as tried-and-tested, practical strategies for workplace survival from other industries.

The author, Bob Hughes, has been a creative for most of his working life: first a calligrapher, then an advertising artist and copywriter before discovering computers and going on to lecture at Oxford Brookes University on the MA in Interactive Media Publishing. He researches and writes about the wider impact of electronics and computers in workplaces world-wide and campaigns on behalf of migrants, refugees and all precarious workers.

Personal Space; Updated,
The Behavioral Basis of Design

by Robert Sommer

ISBN 0954723961

Widely regarded as the classic text on user-centered design of public buildings and spaces, this book studies how people relate to the designed space around them and how the design of that space can affect their behavior.

Its contents are based on comprehensive academic research, the conclusions it draws are highly applicable in real-world projects, and the style of presentation is lucid and interesting. The author; Robert Sommer, deals with topics such as privacy, spatial invasion, small-group ecology and looks at the role of design in settings that include airports, stations, hospitals and schools.

When it was first published this book went through 25 printings in English alone. Although it was written when the study of behavior and design was just beginning to be appreciated, the lessons are still deeply relevant for today's spatial designers and also important to the digital world in the design and analysis of shared digital spaces for co-operative work and multi-player games.

Robert Sommer writes with authority and clarity. He has updated this new edition of his work with an introduction and detailed notes accompanying each chapter describing what has changed in the intervening years, and what has remained the same.

www.ingramcontent.com/pod-product-compliance
Lightning Source LLC
Chambersburg PA
CBHW071157050326
40689CB00011B/2156

* 9 7 8 0 9 5 4 7 2 3 9 7 2 *